Palgrave BioArt

Series Editor

Ionat Zurr, School of Design, University of Western Australia, Crawley, Australia

Reconfiguring living organisms into technologies can change our relationship with the environment, our bodies, and with concepts of materiality, nature, and life itself. What happens when we treat life as a raw material for artistic expressions? *Palgrave Studies in BioArt* presents a series of books written by researchers and artists who manipulate life in scientific laboratories. These artists develop new meanings relating to the concept of life through engaging, provoking, and creating contestable living and semi-living biological artworks. They ask: What is life? What is a body? What are the futures of life? And who is allowed to manipulate life? Such BioArtistic investigations are vital in articulating this new somatic-cultural space. The series will present important and diverse voices discussing frontier biotechnologies and their effects on society, ecology, industry, and life itself. This interdisciplinary series will be of interest to those working in the areas of art and design, science, cultural studies, bioethics, science fiction, and much more. We welcome proposals from researchers and practitioners in the field of BioArt, and cultural/experiential laboratory engagement.

Hyla Willis • Irina Aristarkhova •
Nathanael Elias Mengist • Faith Wilding

Night Sweats: Cyberfeminist Practices

palgrave
macmillan

Hyla Willis
Robert Morris University
Moon Township, PA, USA

Nathanael Elias Mengist
University of Washington
Seattle, WA, USA

Irina Aristarkhova
University of Michigan
Ann Arbor, MI, USA

Faith Wilding
School of the Art Institute of Chicago
Chicago, IL, USA

ISBN 978-3-032-06954-2 ISBN 978-3-032-06955-9 (eBook)
https://doi.org/10.1007/978-3-032-06955-9

Image credit: Hyla Willis
Image Caption: Closeup of subRosa's "Gene Pools" from the *Knowing Bodies* exhibition (2000).

This Palgrave Macmillan imprint is published by the registered company Springer Nature Switzerland AG.
The registered company address is: Gewerbestrasse 11, 6330 Cham, Switzerland.

Acknowledgments

The authors would like to thank, first of all, Ionat Zurr, Bioart Book Series Editor, for her brilliance, wisdom, kindness, and guidance as we embarked on this journey three years ago. We are surrounded by wonderful books in this exciting Series and are glad to be part of it. We would also like to thank Camille Davies, Senior Editor in Cultural Studies, and her team at Palgrave. They have been a pleasure to work with. The Penny W. Stamps School of Art & Design and the Vice-President for Research Office at the University of Michigan awarded us a Catalyst Grant, which enabled us to meet and work on the book together, since access to electronic devices has not always been possible, and we are deeply grateful for that funding, too, especially in this challenging environment for research and creative work. Robert Morris University provided a much-needed sabbatical that enabled Hyla Willis and the team to focus on many aspects of this book's writing and preparation.

We were able to present earlier versions of these chapters at several key venues, benefiting from the collective knowledge and feedback of our professional communities. We thank the FEMeeting: Women in Art, Science, and Technology meetings, hosted in June 2024 by Jennifer Willet, the director of INCUBATOR Lab, professor and a Canada Research Chair in Art, Science, and Technology at the University of Windsor. They selected our presentations for the conference; it was a really special event, and we felt very welcomed. We also thank the New Media Caucus of the

College Art Association, led by Nick Bontrager as its president, for giving us an opportunity to present versions of our texts at New Media Caucus Panel at the CAA conference in New York in February 2025, with Margaret Tan, who attended after a long journey from Singapore. Finally, we thank CAMP: Mumbai-based studio for transdisciplinary media practices, founded by Shaina Anand, Sanjay Bhangar, and Ashok Sukumaran, for the invitation to give a talk in their collectively run artistic space in January 2026. We did our archival research in various locations and especially would like to thank STUDIO for Creative Inquiry (Harrison Apple / Golan Levin / Jim Duesing / Bob Bingham / Marge Meyers) for welcoming us over the years and for hosting subRosa's archive in their library.

Autonomedia, the *Old Boys Network* (OBN) and Cyberfeminist International, Creative Capital, Elena Marchevska, Katy Deepwell's n.paradoxa, and Mindy Seu / Cyberfeminism Index have supported subRosa's projects and their legacy in various ways over the years, and we are grateful to them for that. There would be no Chapter 3 with Virtual Roundtable in this book if not for our generous interviewees and Roundtable participants. Their contributions are inspirational and enrich our book, and their work has been inspirational for us, too: María Fernández, Amelia Jones, Tania Kupczak, Laleh Mehran, Susanna Paasonen, Nell Tenhaaf, and Michelle M. Wright. We thank Margherita Hill, then an undergraduate student at the University of Michigan, who assisted us with images, alt text, and bibliography in the early stages of the manuscript preparation. Finally, we want to acknowledge Lucca, Hyla Willis' ten-year-old dog, who delighted us with her barks and attention during our residency work in Pittsburgh.

In addition to our joint acknowledgements:

I, Hyla Willis, would like to thank Irina Aristarkhova, Faith Wilding, Nat Mengist, Mark O'Connor, all the subRosies, and inspirators of note: Gary Carlough, Katy Deepwell, María Fernández, Raul Moarquech Ferrera-Balanquet, Ryan Griffis, Gita Hashemi, Kathy High, Christina Hung, Adeline Kueh, Steve Kurtz, Elena Marchevska, Guna Nadarajan, Tanja Ostojic, Maria Ptqk, Stephanie Rothenberg, Paige Sarlin, Norie Sato, Lucia Sommer, Susan Squier, Astria Suparak, Marge Tan, Nato Thompson, Wu Tsang, Adam Zaretsky, Ionat Zurr and my mother, Charlotte Writer, RMU colleagues Nadine Englert, Carolina Loyola-Garcia, Ed Karshner, Frank Hartle, Ann Jabro, Heather Pinson, John Sotirakis, Patty Then, Joe Zimmer, and of course, my Union: AFT, Local #3412, have been especially supportive of my time for research and travel. Thank you to the Pittsburgh Lettering Club, the YMCA, Chris, Irina (again and again!), and my beloved "deejay nightmares" friends for life-sustaining and stress-reducing camaraderie.

I, Irina Aristarkhova, am grateful to my co-authors, Hyla Willis, Nat Mengist, and Faith Wilding, for their collaboration, and to Faith Wilding also for a long and profound friendship. Ionat Zurr, the Bioart Book Series Editor, for her substantive feedback and encouragement throughout our work on this manuscript, and for all that she has done for the bioart community, including SymbioticA's pioneering bioart work with Oron Catts. I thank all subRosa members and collaborators over the years who inspired me, especially María Fernández and Elena Marchevska. I thank all my international cyberfeminist sisters from various professional communities, including from the University of Michigan, Penn State University, National University of Singapore, the Society for Literature, Science and the Arts, the Society for Social Studies of Science (4S), the National Women's Studies Association, *Catalyst: Feminism, Theory, Technoscience* journal, FEMeeting, Cultivamos Cultura in Portugal, Hothouse in Singapore and the former CyberfeminClub of St. Petersburg, created by the amazing Alla Mitrofanova and Irina Aktuganova in the mid-1990s. I am grateful to the OBN Group, Ingrid Hoofd, Boryana Rossa, Kathy High, Marta de Menezes, Dalila Honorato, Susan Squier, Lisa Nakamura, Katherine Behar, Margaret Tan, Shirley Soh, Jasmina Tumbas, Nassim Parvin, Anne Pollock, Mindy Seu, and many others who supported me in this work and to whom I am sorry not to list here due to the limitations of space, submission rush, and memory lapses. I thank my dear friends and family members around the world, especially my great niece, Alisa Semenova, who is fourteen now and is probably the most creative member of the family by far, with her drawing and animation practice and social media following. Finally, this book would not be possible without Hyla Willis, my co-writing and co-creating comrade, and a joy to work with.

I, Nat Mengist, would like to first thank Irina Aristarkhova and Hyla Willis for welcoming me into the world of cyberfeminism, and especially Irina, who has been uplifting me for the last six years. I am forever grateful to my first and longest advisor, Phillip Thurtle, and my longtime collaborators on *DIY Homunculi*, Ioan Butiu and Dakota Gearhart. Thank you to Garrett Laroy Johnson, Asa Sakrison, Zach Thomas, Celina Osuna, Joshua Grant-Yount, John Bratt, and Brandon Mechtley for giving me the space to dream weird dreams with the Post-Human Network, and to my friends and colleagues: Michael Beach, Muindi Fanuel Muindi, Thierry Bardini, Mira Petrillo, Anisha Uppugonduri, Gabrielle Benabdallah, Leslie

Coney, and Rukimani PV. Thank you to Stacey Moran, Adam Nocek, Sha Xin Wei, Tyler Fox, Audrey Desjardins, Rebecca Cummins, Stephanie Rothenberg, Beth Kolko, Afroditi Psarra, Charlene Montaño Nolan, and Joshua Scannell for your support; to Kate Kretsinger-Dunham, Kelly Tong, Maza Hailu, and Rebecca Chan for your thought partnership; and to *The Anthropocene Laboratory* editors Fröydi Lazlo and Anna Risell for your dedication. I am also grateful to Cynthia Greenlee, Dorothy Roberts, Alys Weinbaum, Lynn Thomas, and Oliver Rollins for your guidance; to Kathy Woodward, Amanda Swarr, Nassim Parvin, Neda Atanasoski, Bettina Judd, and Filiberto Barajas-López for your mentorship. Special shout-out to my friends Cynthia Anderson, Nick Barr, Estefania Narvaez, Rae Rome, and Marcus Henderson, and to my fellow witches Efrén Cruz Cortés, Margaretha Haughwout, and Sandra Huber. Finally, my partner Bana Hatzey, with whom I do it all.

Faith Wilding would like to thank Hyla, Irina, and Nat (one day we will meet!), all subRosa members and collaborators; my friends over the years; Anat Ebgi Gallery; cherished former colleagues at the School of the Art Institute of Chicago; Vermont College of Fine Arts; the Pembroke Center at Brown University; the Getty Institute for hosting my archive; the many brilliant students and collaborators who have conspired with me along the way, and especially my large and loving family.

Contents

1 **Introduction: subRosa Legacy** 1
Hyla Willis, Irina Aristarkhova,
Nathanael Elias Mengist, Faith Wilding

2 **Terms and Conditions** 31
Hyla Willis

3 **Domain Errors! A Virtual Roundtable** 61
Hyla Willis and Irina Aristarkhova, with contributions
from María Fernández, Amelia G Jones,
Tania Kupczak, Laleh Mehran, Susanna Paasonen,
Nell Tenhaaf, and Michelle M. Wright

4 **Magical Life** 107
Nathanael Elias Mengist and Irina Aristarkhova

5 **Techno-Rant of the Cyborg Elder** 153
Faith Wilding

6 **subRosa Manifestations: A Chronology of Particular Projects, with Notes** 163
Hyla Willis

Index 205

About the Authors

Hyla Willis is an interdisciplinary artist and graphic designer, best known for her work as a longtime member of the subRosa cyberfeminist art collective. She holds a MFA from Carnegie Mellon University, and has been the recipient of two Pennsylvania Council on the Arts Fellowships in New Genres, a Creative Capital grant in Emerging Fields, a MacDowell Colony Fellowship in Interdisciplinary Arts, and an Artist Opportunity Grant from the Greater Pittsburgh Arts Council and was Pittsburgh Center for the Arts' Artist of the Year in 2014 with a solo exhibition "America's Least Livable City," an homage to her home town in California's Sacramento Valley. She teaches Media Arts at Robert Morris University in Moon Township, Pennsylvania. http://www.cyberfeminism.net

Irina Aristarkhova is a writer and scholar who is interested in cultural transformations and new forms of thinking, making, and being. She is the author of *Hospitality of the Matrix: Philosophy, Biomedicine and Culture* (Columbia University Press, 2012) and *Arrested Welcome: Hospitality in Contemporary Art* (University of Minnesota Press, 2020), and editor, most recently, of *UNMYTH: Works and Worlds of Mithu Sen* (Mapin, 2025). She is also a co-editor of *Catalyst: Feminism, Theory, Technoscience Journal*, https://catalystjournal.org. Aristarkhova is Professor at the Penny W. Stamps School of Art & Design and Associate Director at the Digital Studies Institute, University of Michigan, Ann Arbor, where her ongoing research and teaching encompass comparative feminist theory, digital and technoscience studies, and contemporary aesthetics.

Nathanael Elias Mengist is currently a PhD student in Learning Sciences at the University of Washington in Seattle. His collaborative artwork *DIY Homunculi,* featured on the animated video podcast "Life Touching Life," appropriates the politics of alchemy to expand how we imagine both environmental and reproductive justice. Nathanael's research has been supported by the Walter Chapin Simpson Center for the Humanities, and he has published in *Art, Design, & Communication in Higher Education*, the *Companion Publication of the 2023 ACM Designing Interactive Systems Conference*, and *The Anthropocene Laboratory* by 284 Publishing.

Faith Wilding is an independent artist, Professor Emerita, and former Chair of Performance Art at the School of the Art Institute of Chicago. She is a founding member of subRosa, a cyberfeminist art collective. Wilding is a well-known figure in the history of the feminist art movement and was a member of Womanhouse at CalArts in the early 1970s. Her work has been acquired by major museums and exhibited widely at major venues, represented by Anat Ebgi Gallery. Her essay "Where is Feminism in Cyberfeminism?" has been a seminal work for feminism and technology studies. A major retrospective book about her work, *Faith Wilding: Fearful Symmetries,* edited by Shannon R. Stratton, was published in 2018 by Intellect Books.

Abbreviations

AI	Artificial Intelligence
ARTs	Assisted Reproductive Technologies
DIY	Do-it-Yourself
IVF	In-vitro Fertilization
pARTs	Plant-Assisted Reproductive Technologies
SLSA	Society for Literature, Science, and the Arts
STEM	Science, Technology, Engineering, and Mathematics
STS	Science, Technology and Society Studies
UX/UI	User Experience and User Interface Design
US	United States of America

CHAPTER 1

Introduction: subRosa Legacy

Hyla Willis, Irina Aristarkhova,
Nathanael Elias Mengist, Faith Wilding

In this Introduction, we explore feminist contributions to the expanded field of intersections between art, science, and technology. More specifically, we focus on the crossover between feminist contributions to bioart and feminist contributions to technoscience that have come under the banner of cyberfeminism. The questions raised by early cyberfeminists in bioart and in art & technoscience collaborations have only become more prescient in the current moment of gender and sexuality panics, the lies and biases of the artificial intelligence (AI), and the widespread use of assisted reproductive technologies (ARTs) in the context of the continuing presence of reproductive injustices and inequalities. This book, however, is about affirming new creative and life practices that are built on ancient wisdom and collective action, with each chapter offering its own contemporary examples. In doing so, we bring together the legacy of early cyberfeminist art practices and the current debates around creative and communal ways of living life in the age of computational and reproductive technoscience. Throughout the book, we focus on artworks and projects of the cyberfeminist art collective subRosa as a key representative of the importance of early cyberfeminist questions and practices to bioart, tactical media, and technosciences.

H. Willis et al., *Night Sweats: Cyberfeminist Practices*, Palgrave BioArt,
https://doi.org/10.1007/978-3-032-06955-9_1

subRosa's Expanded Cyberfeminist Story

A major contributor to the understanding of intersections of power and technology, subRosa's works raised critical questions about biomedical sciences: examining the flow of capital, tissues and organs around the world; the history of science from intersectional feminist perspectives; questions of power and materiality in STEM (Science, Technology, Engineering, and Mathematics) disciplines; and embodied labor in the age of globalized biotechnical reproduction and extractive life practices of patriarchy and racial capitalism.

Since the early cyberfeminist formations, there have been major cultural, economic, and scientific developments in biotechnology, communications technology, AI, and ARTs. Simultaneously, we witness global attacks and backlash on reproductive, gay, and trans rights and suppression of their history and critical race theory in all levels of education. Those very histories and theories, however, have enabled cyberfeminism to re-enter academic and cultural fields, exemplified by Legacy Russell's *Glitch Feminism* (2020) and Mindy Seu's monumental project *Cyberfeminism Index* (2023) that chronicle cyberfeminist history across a wide spectrum of influences and backgrounds. Contributing to this growing influence of cyberfeminist legacy, subRosa's collaborative practice in bioart and tactical media enables us to revisit the diversity of voices in early cyberfeminist thought and creative practice while tracing specific aesthetic choices and theoretical interventions.

subRosa cyberfeminist art collective was formed in 1998 by a group of artists, scholars, and activists whose lives intersected in Pittsburgh, Pennsylvania. Faith Wilding, in her role as Visiting Artist in the School of Art at Carnegie Mellon University (CMU), had garnered support from the School's STUDIO for Creative Inquiry for an open-ended art research project, "Sex and Gender in the Biotech Century." Initially, this reading-and-discussion group was concerned with the impact on women of advances in ARTs and the sequencing of the human genome. The earliest participants in Pittsburgh included Emily de Araujo, Krista Connerly, Steffi Domike, María Fernández, Camilla Griggers, Maureen Clarke, Christina Hung, Carolina Loyola Garcia, Laleh Mehran, Elizabeth Monoian, Ann Rosenthal, Suzie Silver, Lucia Sommer, Rebecca Vaughan, Faith Wilding, Hyla Willis, and Michelle M. Wright.

This group became a jumping-off point for subRosa, the art collective, which naturally coalesced into a smaller working group of artists as it took on ambitious productions, including art installations, interactive performances, and an edited anthology, *Domain Errors! Cyberfeminist Practices* (2003), edited by Wilding, Fernández, and Wright, and which chronicled some of the very first subRosa art projects by then a core group of six contributors (Domike, Hung, Mehran, Sommers, Wilding, and Willis). Both the large reading group and the smaller art-making group brought a diversity of economic, generational, and cultural backgrounds as well as emerging and highly developed skills in many professions, trades, and social justice movements. This influenced what texts the group read and the interdisciplinary nature of its art output.

The name for the art collective was suggested by Faith Wilding at one of the reading group meetings. It is a Latin phrase meaning "under the rose" with long roots in Christian symbolism and secrecy, which Wilding was familiar with, having been raised in a German Anabaptist colony in Paraguay. Despite being a heterogeneous group when it came to religious backgrounds and practices, group members liked the "floral anonymity" and the fact that there were so many notable women with the names Rosa (e.g., Parks, Luxemburg, Bonheur), Rosalyn (Franklin), Rosie (the Riveter), etc. subRosa is a term also used in spycraft. The camelcase capitalization was adopted to reflect digital file naming conventions. It is also close in spelling to the Spanish word for "tasty." With time, it became amusing to the group members to witness "subrosa" circulate as a name for other companies, such as a Croatian car rental company and a BMX bike manufacturer popular with teenage boys.

The surrounding years witnessed the popularization of In-vitro Fertilization (IVF) , part of the growing ART industry; the growth of the World Wide Web, smartphones, Amazon, Google, and of Dolly, the first mammal cloned from adult stem cells. The 9/11 attacks bookended by high-tech Gulf Wars, further shaped the shared discourse and personal experiences of the members. Following the 2002 publication of the *Domain Errors! Cyberfeminist Practices*, subRosa members dispersed throughout the country, returning to Pittsburgh for the intense work sessions that laid the groundwork for performances, installations, and publications.

After its formation, subRosa quickly became part of the pioneering cyberfeminist art conversations of *VNS Matrix*, the *Old Boys Network* (OBN), tactical media projects such as *RTMark* and *Critical Art Ensemble*, and art and biology communities. Various subRosa members knew, collaborated with, and overlapped with these groups. *The Institute for Applied*

Autonomy emerged from Pittsburgh the same year and frequented many of the same spaces while rubbing up against the extreme technophilic atmosphere. It was an exciting time of transition in Pittsburgh, in particular at CMU's School of Art. Artists Paul Vanouse, Bryan Rogers, Nell Tenhaaf, Stelarc, Simon Penny, Critical Art Ensemble members, Hope Kurtz, Steve Kurtz, and (at the time) newer faculty Jim Duesing and Suzie Silver were part of critical discussions and socializing with varying degrees of overlap. Institutional funding enabled some subRosa members to present at the *Performative Sites* conference at Penn State University, 2000, and *The Next Five Minutes: A Festival of Tactical Media* (N5M) conferences in Amsterdam, in 1999 and 2003. There, subRosa met with other tactical media scholars, activists, and artists, among them Raqs Media Collective members and Irina Aristarkhova, who at the time was starting her *Cyberarts and Cyberculture Research Initiative* (CCRI) at the National University in Singapore (Yap and Han 2021). Their meeting led to the formation of *Undercurrents: A List-Serve about how cyberfeminism, new technologies, postcoloniality, and globalization are interrelated* (see Sect. 1.3) and a 2003 N5M Tactical Media Lab *Matrixial Technologies* collaboration between subRosa and Aristarkhova, Margaret Tan, and Adeline Kueh. *Matrixial Technologies* was hosted jointly by CCRI at the National University of Singapore and by Lasalle College of the Arts (Yamin and Yap 2023, Yap and Han 2021, subRosa, Aristarkhova, Tan, and Kueh 2003). The following year (2004), subRosa went on to participate in "*Bio-Difference: The Political Ecology*," Biennale of Electronic Arts Perth, at the Lawrence Wilson Art Gallery, the University of Western Australia, where, at some point, they were also hosted by Ionat Zurr and Oron Catts of SymbioticA. At all these events, subRosa circulated copies of the founding manifesto (subRosa 1998), enclosed in biohazard bags (see Chap. 6). Every member of the group contributed from their years of experience with writing, copyediting, typesetting, video production, drawing, painting, photography, filmmaking, radio plays, dance, experimental music, project management, cooking, teaching, and grassroots organizing.

Cyberfeminism and Biotechnology

Why this book *now*? We have come to work together on this book for two main reasons. First, we wanted to document subRosa's practice because it represents an important point in the development of critical and creative cyberfeminist approaches to technoscience (as evidenced by articles, chapters, theses, scholarly works, and various archives related to subRosa's work, among them are Falconer 2013, Jordan-Young 2010–11, Marchevska and Walkerdine 2016, Motter 2011, Ptqk 2017, Timeto 2010, Faith Wilding Getty Archives, Steffi Domike University of Pittsburgh Archives, and subRosa's materials at the STUDIO for Creative Inquiry at CMU). While there are important books surveying cyberfeminism and cyberfeminist art (e.g., mentioned above Russell 2020 and Seu 2023, among others), we need more books about important individual art groups and contributors to cyberfeminist technoscience. This series provides an excellent platform for that task. Second, we wanted to show the timeliness of these works in the current moment and their relevance to and legacy for new questions, as we use our discussions about subRosa's works to contribute to the ongoing debates around computational and reproductive technologies and the worldbuilding that finds forms to refuse capture and extraction in creative and nourishing ways.

subRosa's works at art festivals, exhibitions, on-campus and public performances, academic conferences, and dissemination of writing exhibited frustration with the dearth of feminist conversations critical of emerging technologies, be they what was then called "information and communication technologies" (ICTs) or biomedical and reproductive technologies. Thus, in 2003, during their residency in Singapore (that was investing in life sciences and their implementation in research, industry, and healthcare settings), collaborators there provided subRosa with a chance to visit many sites of cutting-edge research in ARTs and bioengineering, including stops at laboratories and participation in seminars and lectures. subRosa visited endangered species cloning lab at the National University Hospital; CordLife, a licensed lab and storage bank for preserving and extracting umbilical cord blood stem cells (by now, 2025, it has become the largest in Asia), where they interviewed the chief scientific researcher and a laboratory manager. subRosa participated in a seminar "Art meets Bio-Science: Methodology and Ethics of Interdisciplinary Work" at the National University of Singapore, with a subsequent public discussion about how artists and scientists might approach interdisciplinary projects that address public issues in cloning, genetic engineering, stem cell research, and what the ethics and aesthetics of such partnerships or collaborations might be. They conducted a two-day workshop "Cloning Cultures" at Lasalle College of the Arts, addressing ideas about becoming post-human and entering the clonal age of genetic engineering and body enhancement by teaching the basic processes of cloning technologies and how stem cell lines are derived (see Fig. 1.1). subRosa also presented a public lecture "Art, Activism, and Biotechnology" as part of the Lasalle's "Art and Technology Talks" (another visitor to the series, Guy Ben-Ari of SymbioticA, gave his lecture during the same time, after which subRosa and Ben-Ari had a public discussion). Importantly to their topic of interest in ARTs, subRosa had a chance to visit and tour KK Women's Hospital where most births take place in Singapore, including the Neonatal Intensive Care Unit (NICU) for premature infants, and the maternal and obstetrical wards and facilities, where they had a two-hour meeting with doctors who answered questions about fertility, pregnancy, birth, and post-natal treatment and care in Singapore (where subRosa members noted that the hospital catered to varying cultural and religious groups' needs in relation to birthing practices, when requested by birthing persons themselves, coming predominantly from the Chinese, Malay, Indian, and Eurasian populations of Singapore). And subRosa members also engaged with women's rights and women artists' organizations in the country.

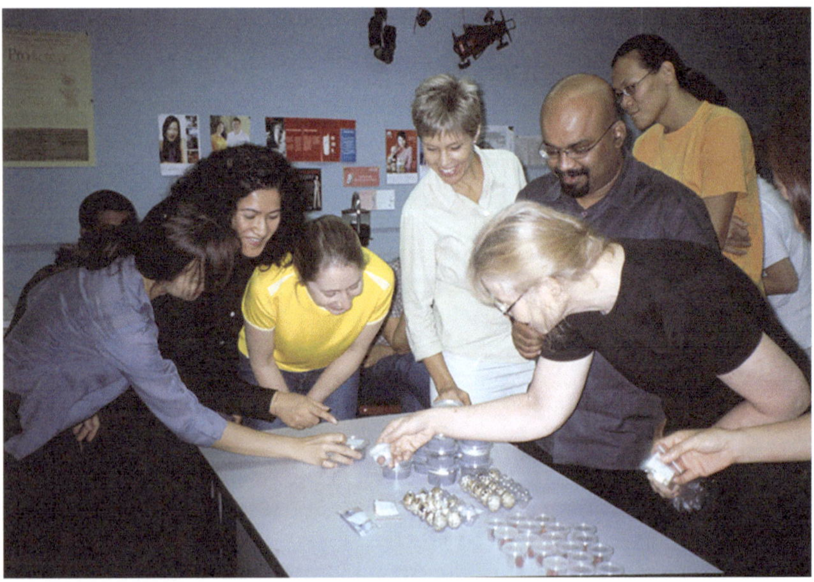

Fig. 1.1 Image from subRosa's workshop at Lasalle College of the Arts (Singapore, 2003).

The thematic ways in which biotechnology and the fertility industry held particular impact on women and non-binary persons made fruitful ground for subRosa's feminist artistic praxis, as their work often appeared alongside different kinds of bioart, including those which primarily treated biology as an expanded art material to experiment with. This new aesthetic ecology prominently included the body. For those of us who were part of establishing and growing the new media art community in the 1990s, we included embodiment in the triad of "hardware, software, and wetware." Wetware was the body and life; connecting tissue culture with computational media. It is within this understanding that subRosa was introduced as "a Cyber Feminist Cell of Cultural Producers Working with Wetware" in the volume *Tactical Biopolitics: Art, Activism, and Technoscience,* edited by Beatriz da Costa and Kavita Philip (2008, VI).

These three terms—hardware, software, and wetware—allowed for the encompassing of various configurations toward a variety of aesthetic expressions: for some, this was the time to experiment with new materials, such as tissue culture, or body-machine entanglements (e.g., Stelarc and SymbioticA); for others, this was the time to use the new media to highlight old prejudices or inequalities (e.g., the Obabikes and Fusco discussed below). Some approached new media, including biotechnologies, with fascination; others—with critique, irony, or caution. Many, with both fascination and caution. And others, furthermore, employed "cyber" toward activism. With time, environmental concerns and care about all living beings has become an important question in bioart, and for some, it was an important issue from the beginning (see, for example, the excellent earlier collection *The aesthetics of care?: the artistic, social and scientific implications of the use of biological/medical technologies for artistic purposes,* Catts, 2002).

In several of subRosa's projects, dry lab equipment was demonstrated by performers. For example, the real working cannulas and other zygote-collection tools were shown at *Expo EmmaGenics* (2001). In other cases, the audience could learn how to extract DNA from strawberries (*Bodies Unlimited* workshop, 2010); collect DNA from the audience on cheek swabs (*Sex and Gender Education Show,* 2000/2002); practice streaking cells onto agar-lined Petri dishes; see an experiment in cloning bacteria; and learn DIY cell culturing such as yogurt making (*Epidermic! DIY Cell Lab,* 2005, see Fig. 1.2).

In other projects, the lab setting was more ironic, traversing into what we discuss below in relation to a tactical media aesthetic (see Fig. 1.3).

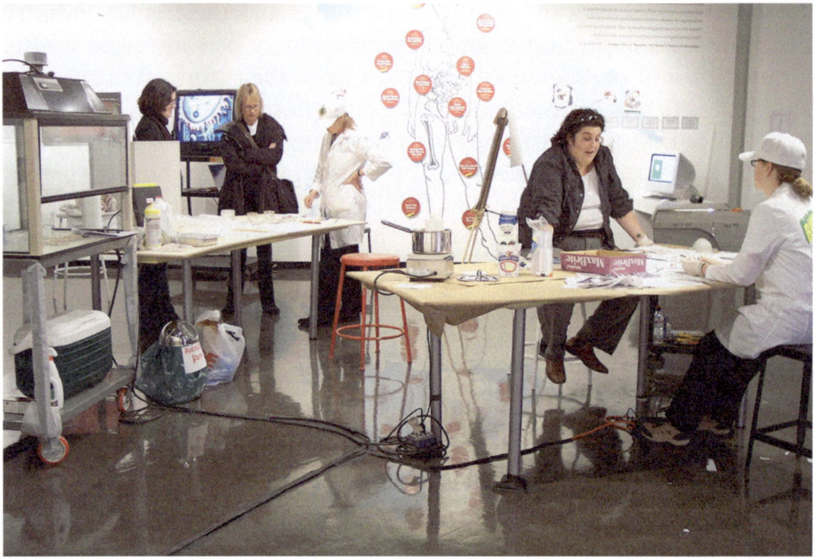

Fig. 1.2 Installation view from *Epidermic! DIY Cell Lab*. subRosa, Betty Rymer Gallery, 2005. At the table in the foreground, participants could make collages from scientific and alchemical images and learn how to culture yogurt cells at home. In the background, there is a table for learning how to streak petri dishes or observe a scientific experiment in cloning cells.

For several subRosa members, living matter continued to be an important site of intervention and manipulation in their individual professional work. Thus, Faith Wilding taught a course on how to grow yeast at the School of the Art Institute of Chicago in the Performance Art Department in the early 2000s. And in 2005, Christina Hung created *All Things Grow With Love*, (Hung 2005) part of her *Serratia Marcescens* series, in which she stamped imagery onto agar-lined Petri dishes and allowed the red bacteria to reveal imagery with their natural growth.

At the time, when reproductive technologies were addressed in early bioart and new media art communities, it was often through a speculative future lens of what would become possible, not necessarily focusing on questions of their eugenic tendencies or capitalist and patriarchal extractive practices. For example, the Ars Electronica Festival in 2000, where the theme was "Next Sex: Sex in the Age of Reproductive Superfluousness"

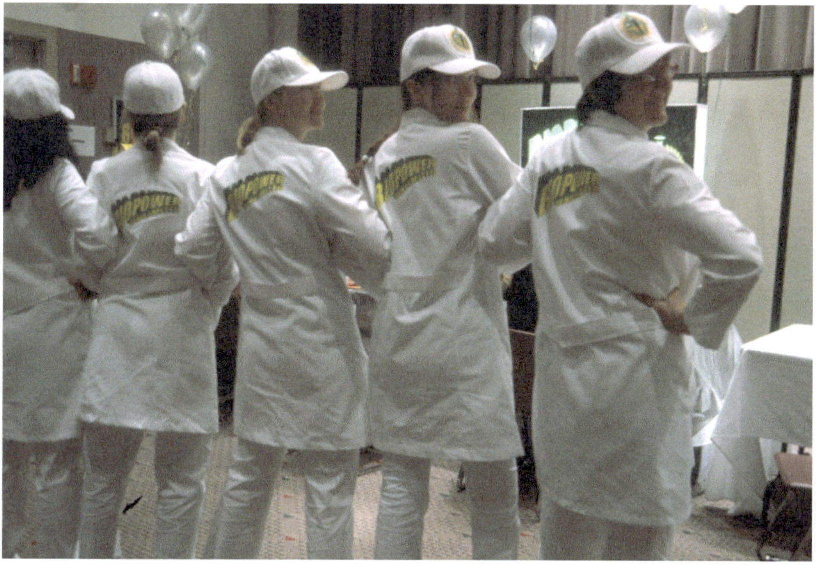

Fig. 1.3 subRosa members lined up in lab coat costumes for the *Biopower Unlimited!* campus intervention created for Bowling Green State University's New Media & Art Festival. L to R from behind are Laleh Mehran, Hyla Willis, Faith Wilding, Lucia Sommer, and Steffi Domike (subRosa, 2002).

(Stocker 2000) leaned in heavily into the superfluousness of the title. The festival showcased biomedical and bioengineering research in artificial wombs and other reproductive biotech with the lightness and brightness of a trade show but without subRosa's underlying biting feminist and anti-eugenicist critique. Here, seemingly the same aesthetics of a science education show served speculative reproductive futurism, as if asking its audience: 'When you have an artificial womb, who cares about reproductive justice?'(see Chap. 4).

The fact remains that practices and definitions of ARTs continue to be traveling "in and out" of birthing bodies. As Aristarkhova predicted in *Hospitality of the Matrix* (2012)—built on ideas that also fed the "Matrixial Technologies" collaboration with subRosa in Singapore—ectogenetic desire continues to seek to erase the experience and labor of birthing and nursing. It is enacted through a downplaying of the enormous labor of birthing, nursing, and maintaining life through machinic and xeno-mimicry, truncation, and attempts at substitution of oviparity. Today, ectogenetic desire is expanding into the tropes of "artificial" and "autonomous" everything. We are promised that AI will soon create new worlds and merge with Life itself, turning from a helping hand into "generative" and "agential" modes of being and becoming (see Chap. 2). In this prosthetic context, renewed retrograde politics and hysteria around transmen's pregnancies, non-binary bodies, and sexualities, and the erosion of reproductive rights, are subsumed in biomedicine and bioethics by expensive, mostly stalled and curiosity-driven, research in artificial placentas, wombs, and xenosurrogacy. Projects that work on biomedical and technoscientific substitutions and enhancements of life obscure less expensive solutions that could be implemented in the present and do not translate into or prompt research and public policies on widening access to ARTs or covering the cost of sustaining life: childcare, healthcare, food, housing, mobility, or education.

Cyberfeminist Intersections

subRosa members' emphasis on the histories of civil rights and labor movements enabled them to widen the scope of their predominantly educated, white, cisgender community in Pittsburgh. Echoing Legacy Russell's more recent critique of cyberfeminism's blindness to racism (Russell 2020, see also Chap. 3), individual subRosa members and their collaborators were involved in initiatives that brought more robust discussions around race, colonialism, and feminism into cyberfeminism, bioart, new media, and science art. Thus, in early 2002, María Fernández, Faith Wilding, Coco Fusco, and Irina Aristarkhova launched a list-serve on the thing.net platform called "Undercurrents: A List-Serve about how cyberfeminism, new technologies, postcoloniality, and globalization are interrelated." This combination of words was very important to them as moderators, and they invited to the list artists, activists, and scholars who had already been addressing these intersections in their own communities and whom they met at various venues. Participants were based in South Asia, Southeast Asia, North America, Australia, Central and South America, Africa, and Europe.

The list originated from the Undercurrents moderators' frustration with the premise, often championed at the time, that new technoscience and digital media would erase borders and overcome inequalities in identities, creating new digital commons. In a welcoming message, the moderators wrote:

> As much as we support the democratic goals of many who have contributed to alternative discourses within net.culture, we do not agree that the ideal of a digital commons, feminist or otherwise, necessarily transcends the problematic logic of race and racism. We are deeply skeptical of such assumptions because we understand that race and racism involve much more than skin, bodies, overt segregation, or physical violence. We argue instead that race is manifest in both the essentializing ventures of law and science and in the arenas of performativity that denaturalize and de-essentialize embodiment, including cyberspace. (Fernández, Wilding, Fusco, and Aristarkhova, quoted in Fernández, Fusco, and Aristarkhova 2002, n.p.)

Adding the importance of postcolonial theoretical ideas, Fernández, an Undercurrents moderator, art historian, and a frequent collaborator of and contributor to subRosa, argued that:

> [T]he lack of engagement with postcolonial studies constitutes a major gap in theories of new media. This gap is even greater in cyberfeminism, a formation that until recently denied having links to feminism. Often, I am asked what relevance postcolonial studies have to cyberfeminism. That this question should be posed in an era of permanent warfare, seemingly unending migrations, racist national security and immigration policies, brutal nationalisms, global ecological and economic crisis, anti-globalization movements and renewed and continuous efforts in various parts of the world to exploit and control women's bodies and labor invariably surprises me. Are these phenomena conceived as independent from the complex of relations, transactions and interventions facilitated by electronic technologies? Are they regarded as irrelevant to women? (Fernández quoted in Fernández, Fusco, and Aristarkhova 2002, n.p.)

The moderators also wanted to unsettle the Western hegemony of new media and science studies that often-portrayed race and colonialism as "low tech" and irrelevant to cyberspace or bioart, where "anyone could be a dog," a pixel or a cell. The topics covered by the list were wide-ranging. Fusco recounted that among the topics raised in the Undercurrents were:

> the misuse of the term essentialism as a tactic for denying the validity of experience, the meaning and influence of technoformalism; the relationship among whiteness, the celebration of disembodiment, and cybertheory;

> the problematic association of multiculturalism with ethnic fundamentalism; the viability of tactical media, the role of testimony in the understanding and interpretation of state violence around the world, the contributions of afrofuturist thinkers to debates about new technologies, and more. (Fusco, quoted in Fernández, Fusco, and Aristarkhova 2002, n.p.)

This was the context of a cyberfeminist art group at the end of the twentieth and early twenty-first century. subRosa quickly became a highly distinctive art collective addressing the topics that very few new media, bioart, or tactical media collectives focused on at that time: reproductive justice through the lens of the *body as full of mediated products that can be extracted and exchanged, bought and sold, within research and on the open global market.* Feminist histories of technoscience and art enabled subRosa to chart a unique path within new media and tactical media art, focusing on these biomedical extraction practices within global markets of gender and racial capitalism with regards to the body with organs, the provenance of tissue culture itself, and the cost of emotional, physical, and reproductive labour that the body is subjected to due to various kinds of desires, fears, hopes, and capital markets that manipulate them. In addition to tactical media and bioart, subRosa's cyberfemnist art works have encompassed net-art, activist theatre and performance, Soviet constructivism, Brechtian interventions, feminist art history, and political art from the civil rights movement, adding to the varied aesthetic of cyberfeminist contributions (Seu 2023).

SmartMom

In this section, our case study of one of subRosa's earliest works, *SmartMom*, will open an analysis of cyberfeminist connections to tactical media as it heralds its contribution to tactical media theory and biopolitics that is explored in the next section. Willis and Wilding started working on *SmartMom* in the late 90s. They completed the work in 1998, around the time of subRosa's formation, and *SmartMom* became known as the first subRosa web-based project. It was exhibited on many occasions, with one notable location being the Reina Sofia Museum in Madrid, Spain (2005), as part of the travelling exhibition "Violencia sin Cuerpos."

SmartMom was inspired by CMU's context as being a prime location for military research that benefited from government contracts. subRosa took *the Smart T-Shirt,* which was originally developed at Georgia

Tech (similarly technoscience-oriented university), for DARPA (Defense Advanced Research Projects Agency) as a blueprint for a cyberfeminist project. The concept of the *DARPA T-Shirt technology*, as presented at the time on DARPA's website and no longer available, created a notion of the "warfighter" for the new technological age. subRosa connected this early wearable tech with the concept of the Cyborg, also originally developed for the military, by Manfred Clynes and Nathan Kline, and presented in their talk and then article from 1960, "Cyborgs and Space" (Clynes and Kline 1960). By the time of the development of *SmartMom*, the concept of the Cyborg had already been appropriated and popularized by Donna Haraway for her 1985 "Manifesto for Cyborgs," an inspiration for cyberfeminism and cyborg studies, including for subRosa (Haraway 1991).

subRosa researched DARPA's website description of the *Smart T-Shirt*, which was supposed to include "a chip that will create an artificial human immune system; smart fabrics that self-clean and self-decontaminate; and a host of new biomedical tools for "maintaining combat performance" through various biotech enhancements of the neural system, injury repair, genetically modified digestive bacteria, and much more" (subRosa 2008, 240). subRosa reported that soon after they quoted the DARPA site in their *SmartMom* project, "the DARPA Web page on the *Smart T-Shirt* was taken down for unknown reasons" (subRosa 2008, 240). These promises, ironically, also unrealized, echo the original visions of Clynes and Kline of not only inventing technology to maintain a human body but also a technology to alter it, in order to suit war conditions or outer space needs.

Satirically appropriating the language of a "grant proposal" (to submit to DARPA, perhaps?) to develop their own *SmartMom* wearable innovation, subRosa's proposal followed the structure of presenting the problem and offering a solution. The problem "identified" by subRosa was that women continued to resist proposed cyborg adaptations. This added to a long history of women's resistance to attempts at controlling them and their bodies, especially during pregnancy, being "the problem":

> Historically, women's bodies have been notoriously resistant to machine adaptation or medical regulation. The unpredictable ebb and flow of menstrual cycles, hormones, moods, libido, weight loss or gain, metabolism, ovulation, pregnancy, gestation period, fertility, and natural birth rhythms, have severely tested scientific control and management methods. … Using human germ cells manipulated in the laboratory, reproductive scientists are now able to create genetically engineered embryos to implant into human females. But the pregnancy and birth processes are still far less controllable

> although new methods are continually being tested. ... While doctors try to regulate the lives, activities, and diets of their patients, women tend to be resistant to this form of control and many of them habitually disobey doctor's orders and lie about what they have been up to. Add to that the spread of the practice of using surrogate mothers by infertile or older couples, women with health problems, gay couples, single men, and others. Increasingly, those who hire surrogate mothers are seeking the legal right to monitor and prescribe their lifestyles, diets, and activities. But how is this to be done without physically confining the women, or having her followed at all times? Indeed, with declining birth and fertility rates, it is in the interest of all citizens to assist in the surveillance and protection of all pregnancies! Until now, doctors have lacked a foolproof and objective way of constantly monitoring their remote patients, as well as a way of treating them if they cannot be there physically. (subRosa 2008, 226)

What a calamity! subRosa ironically implied here, exaggerating the "problem" in order to then provide their own "solution." While DARPA focused on warfare and the welfare or efficiency of their soldiers and visions of the cyborg in the universe, in reality, most contemporary wearable applications have come to the market as body cycle monitors (sleep, exercise, etc.) and surveillance systems, rather than the above-promised transformation of "a chip that will create an artificial human immune system; smart fabrics that self-clean and self-decontaminate," or "injury repair" in real time by wearing the *Smart T-Shirt*. Thus, the *SmartMom* project highlighted the failure of priorities, when inordinate amounts of money and man-hours (sic!) are spent on warfare and then, hopefully, trickle down into society to make money out of citizens' bodies trying to monitor and maintain themselves. The absurdity of the *SmartMom* solution reinforces this wasteful logic:

> With a combination of the rapid advances in biotechnology, genetic engineering, and smart technology, it seems that at last the tools are at hand to make it possible to include women in the central project of a new kind of technically engineered and assisted biological evolution, which holds out the hope of the birth of a new race of cyborg platforms and organisms. While the military will at first be the chief beneficiary of this technology, it also has immediate and far-reaching benefits and applications in civilian society. It is now possible to control and manipulate reproducing women because new technologies enable the surveillance of women's 'natural' pregnancy and birth processes through telepresent obstetrical monitoring and intervention. Henceforth reproducing women will be able to live and give birth

> in, technological, machinic, or other environments such as space capsules, extraterrestrial environments, remote battlefields, dangerous urban areas, remote rural places, nuclear submarines, and the like, without endangering their offspring, and without altering the biological heredity of their embryonic organic platform which has been genetically engineered to adapt to these environments. Coupled with advanced reproductive technologies that can also be delivered telepresently by Smart technologies, the new remote pregnancy and birth monitoring and manipulation systems represent a breakthrough in cyborg reproduction. Further, the research also holds the promise of complete telepresent monitoring of surrogate mothers who can be systemically manipulated through Smart biotechnology and telepresent supply and control systems. (subRosa 2008, 226–7)

Notice the "official" sounding language here. The SmartMom proposal, as it would be in the "real-world" scientific proposal, came with several concept drawings and sketches for future prototype and development (see Fig. 1.4). In one drawing, we see a woman connected to the above-mentioned capsule. On another, we see a kind of a vest, that subRosa also called the Sensate Pregnancy Dress, a "nifty item that uses optical sensors connected to a web of coded fiber-optic lines leading to a radio transmitter provides constant monitoring of body systems and data such as heartbeat, blood pressure, fluid levels, nervous functioning, the mother's fantasy life, sexual and eating urges, and the like" (subRosa 2008, 227).

In their writing about this cyberfeminist appropriation of military technology, subRosa cited the Situationist concept of "detournement," altering the original meaning of the text and images and, in this process, revealing their hidden, always-already there, "intended," usage. This detournement altered the original meaning behind the military idea of a smart soldier in his Smart T-Shirt, which has revealed the disturbing and ridiculous nature of the original DARPA proposal. In *SmartMom* and other works, subRosa used typography and graphics to mimic and ironically reframe appropriated military emblems and arms (see Fig. 1.5).

Rather than "turning the tables" by, for example, offering to surveil the military and the powerful with their own smart t-shirt, this work provides biting and revealing satire. subRosa's *SmartMom* satire puts it in one group with other feminist and intersectional works of the time, notably, net-art works by Mendi+Keith Obadike (https://obadike.squarespace.com). In those works, they addressed the big questions about who matters, whose life matters, in new technological and scientific environments, and how new they really are when it comes to trafficking

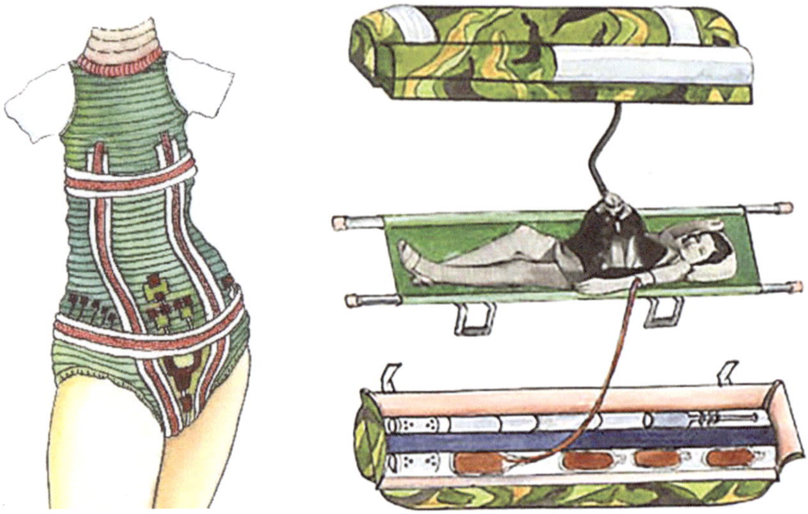

Fig. 1.4 Two images of subRosa's *SmartMom* digital art project. On the left, a watercolor and pencil drawing depicts a concept of *SmartMom* torso shirt with a chip with wires, and on the right, a watercolor and ink drawing with collage from *SmartMom* digital art project depicts a pregnant woman in a capsule being connected via wires to the top cover of the capsule for monitoring, surveillance, and "for her own safety." These drawings use the Situationist tactic of "detournment" to satirically appropriate the concept of Smart T-Shirt military technology developed by DARPA (U.S. military research department) for remote battlefield medicine (subRosa 1998, redesigned in 2000 & 2009).

in anti-Blackness (see, for example, Simone Browne's discussion of this work in Browne 2015 and 2017). In the work titled *Blackness for Sale*, the Obadikes let the buyers "make bids" to purchase "Keith Obadike's Blackness" on the eBay platform on August 8, 2001. In their adoption of the eBay platform, the Obadikes revealed the historical presence of anti-Blackness as a data point, an argument theorized by Jessica Marie Johnson in "Markup Bodies: Black [Life] Studies and Slavery [Death] Studies at the Digital Crossroads" (Johnson 2018). Ebay took the work down a few days later, "flagged" as inappropriate, but not before a few bids were obtained (Browne 2017).

In a similarly reappropriative satire, *SmartMom* highlighted the underlying power structures and the impulse to surveil the pregnant and birthing

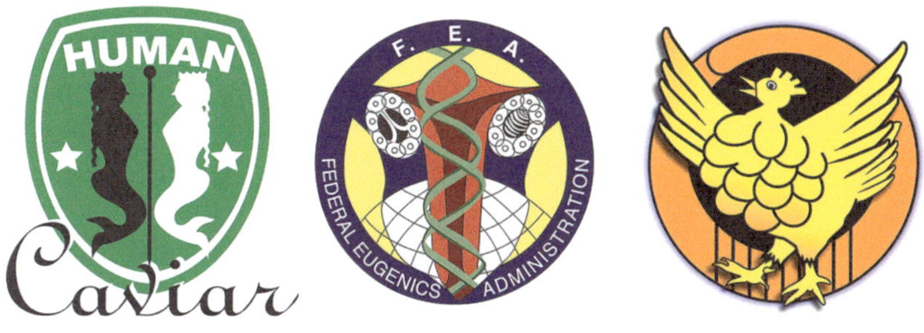

Fig. 1.5 A sampling of corporate-, government-, and military-style emblems created by subRosa members for use in tactical media projects (subRosa, 2002–2004).

person by cheekily proposing an "innovation" to make control over the generative and reproductive power of pregnant persons more efficient, so as to complete the capture of and profit from the generative body. Appropriating an image of a model, a smiling woman posing probably for a magazine from a supposedly bygone time (1940s? 50s? 60s?) lightens up the shocking reaction; to make this reference to a total capture of a refusing person seem digestible by offering a cushion of irony that might not be as helpful today as it seemed in 1998.

The pregnant woman in the image is an "extra white" middle-class model as she also fits into a nuclear family aspiration of the post-war US with its privatized parenting expectations of a stay-at-home mom who has time to get into this capsule to be surveilled—monitored—"guided." She is privatized by the family structure, which she is supposed to aspire to (she is pregnant, after all, and we are to assume, married, otherwise she would not be on the cover of any publication?), and by the *SmartMom* machine and its biotech owners (whether with consent or not remains unclear in subRosa's above-quoted *SmartMom Proposal*). What is clear is that her data is no longer "hers." Placenta-like connections to the top and lower covers of the capsule do not make her look smart. If anyone or anything is smart here, it is the machine. She does look like an embryo herself. Instead of trying to create an artificial placenta or womb as in our earlier discussion on ectogenetic desire, here we see the whole person being encapsulated by and connected to an incubator (as in the *Matrix* movie). The capsule itself follows the military pattern of camouflage, a blend of a combat apparatus, a bariatric chamber, and a (more?) gender-appropriate tanning bed.

SmartMom represents a scandalous (one hopes?) "solution" to pregnant persons' "ongoing resistance" (see Wilding and Willis 2016). Throughout the book, we come back to this issue of an omnipresent drive to capture and privatize the living through the expectation of individual options and decision trees, and the creative and collective paths of refusal. And it is these various degrees of proximity to and involvement in feminist, labor, and civil rights movements that position subRosa's cyberfeminism as central in our theorizing of bioart and tactical media anew.

Tactics of the "Weakened"

While the influence on cyberfeminism and subRosa by major figures such as Donna Haraway, for example, is more obvious and frequently mentioned, here we want to address other influences, namely, Michel de Certeau's notions of the "tactical" and the "weak" as presented in his book *The Practice of Everyday Life* (1984) and the concept of "tactical biopolitics" that Beatriz da Costa and Kavita Philip proposed in their book *Tactical Biopolitics: Art, Activism, and Technoscience* (2008). Both books have been influential for many tactical media theorists and bioartists. We first analyze their concepts here and especially challenge de Certeau's ideas in ways that are different from most other critiques of the tactical and tactical media.

Michel de Certeau proposed to focus on resistance practices of everyday life as a counterpoint to authoritarian tendencies of power, including in the institution of science. His ideas resonated with Foucault's concepts of care of the self and biopolitics, as well as Situationists' attention to reappropriation (e.g., above-mentioned *detournement*) and minor interventions into the apparatuses and *dispositif* of power, with their institutionalized, calculated, and possessive strategies. Art has been an important site of alternatives to power, including the power of science and technology, for many thinkers of the time, including the Frankfurt School, Frantz Fanon, and Sylvia Wynter. Art represents an important site of shifts in various kinds of paradigms and experimentation with forms that are to be materialized in the future on a much larger scale, as exemplified by *SmartMom*.

This hope for a much larger, future scale of resistance and impact is seen in the definition of Tactical Media by Critical Art Ensemble, subRosa's early collaborators and co-conspirators at the time, and which was closely aligned with subRosa's own ethos as they developed the group: "Tactical media … encourages the use of any media that will engage a particular socio-political context in order to create molecular interventions and semiotic shocks that collectively could diminish the rising intensity of authoritarian culture" (*Critical Art Ensemble* 2001, 10).

There are a few points of affinity between this definition and tactical media's original definition borrowed from de Certeau's commitment to tactics as the "weapons of the weak." Thus, in their *ABCs of Tactical Media* (1997) Garcia and Lovink define tactical media as:

> what happens when the cheap "do it yourself" media, made possible by the revolution in consumer electronics and expanded forms of distribution [. . .] are exploited by groups and individuals who feel aggrieved by or excluded from the wider culture. Tactical media do not just report events, as they are never impartial; they always participate, and it is this that more than anything separates them from mainstream media. [. . .] Tactical media are media of crisis, criticism, and opposition. This is both the source of their power, [. . .], and also their limitation. (Garcia and Lovink, quoted in da Costa and Philip, 2008, XVII)

We return to the final point on limitation after outlining several of de Certeau's positions on who is considered strong and who is weak in his original research on the tactics of resistance.

For de Certeau, who borrowed this war vocabulary from Carl von Clausewitz, the strong take over land and power by using their "strategies" of the law, the military, and establishing other institutions that help them maintain their power. The weak, in their turn, invent "tactics" from below that often proliferate and undermine the strong: "We are concerned with battles or games between the strong and the weak, and with the 'actions' which remain possible for the latter" (de Certeau 1984, 34). Problematically presenting these battles as a kind of a game, de Certeau's examples of the tactics included walking / loitering in a city, as pedestrian resistance to urban organization and construction of the strong; the colonized's resistance to the imposed customs, food, religion, and knowledge systems by secretly maintaining their own; and artists' tactics of continuing to make work and invest time in works that provide little ability to be managed or become useful and efficient to authority. Engaging with Foucault's notions of biopolitics, among others, de Certeau also used "tactics" as an umbrella term for any resistance of those who seek freedom and justice from those who exercise their power through direct military and physical force and violence or other strategies of the powerful—for example, during colonialist violence, political totalitarianism, or political oppression.

De Certeau defined strategies and tactics as follows: “I call a strategy the calculation (or manipulation) of power relationships that becomes possible as soon as a subject with will and power (a business, an army, a city, a scientific institution) can be isolated” (1984, 36). Thus, as in management, every “strategic” rationalization “seeks first of all to distinguish its ‘own’ place, that is, the place of its own power and will, from an ‘environment’ ” (ibid).

Relevant for us here is that de Certeau placed science on the side of the powerful, since knowledge is one of the strategies that the powerful can use to accumulate more space and resources to disempower others. This critique of science as being on the side of the strong is directly applicable to cyberfeminist tactics of approaching biotechnology and science. If not based on “scientific knowledge,” then what do “the weak” use? They are based on “wisdom,” for de Certeau, as knowledge from below, that takes time to accumulate through experience, trial, and error. These tactics are temporal, wisely choosing what is best to employ and where best to strike in this very moment. Science here is aligned with military strategy and politics, directly impacting what is to become tactical biopolitics of the Other (that becomes the weak): “Strategies pin their hopes on the resistance that the establishment of a place offers to the erosion of time; tactics on a clever utilization of time, of the opportunities it presents and also of the play that it introduces into the foundations of power (de Certeau 1984, 39).

In *The Rant of the Menopausal Cyborg* (2003), Faith Wilding called for various kinds of tactics of resistance and refusal when she wrote:

> Revel in your uselessness; don’t work; demand full service. Resist the imperative to retool or retrain. Refuse the compulsion to be wired, but get online and mouth off if you want; refuse to be smart, new, and improved. Make coalitions with younger people to help you take over the nursing homes and convert them into livable homes and residential collectives. (Wilding 2003)

Writing about the ubiquitous nature of contemporary power structures and their strategies in modern times of the post-WWII global changes, de Certeau also considered what happens when the “weak” get unmoored from their tactical traditional wisdom and no longer have a place to hide:

> Tactics are more and more frequently going off their tracks. Cut loose from the traditional communities that circumscribed their functioning, they have begun to wander everywhere in a space which is becoming at once more homogeneous and more extensive. Consumers are transformed into

> immigrants. The system in which they move about is too vast to be able to fix them in one place, but too constraining for them ever to be able to escape from it and go into exile elsewhere. There is no longer an elsewhere. (de Certeau 1984, 40)

As if anticipating the digital revolution and proliferation of life surveillance, de Certeau sees this absence of any other "elsewhere" in transformations that will come with the rise of a "cybernetic society." Therefore, in this close reading of de Certeau, we can find the conundrum of the tactical. One could speculate that subRosa's *SmartMom* responds to this wetware proliferation of power and control, where nothing else is left but tactical witnessing. In the language anticipating Deleuze and Guattari, de Certeau weaves a familiar rhetoric of no escape:

> the "strategic" model is also transformed, as if defeated by its own success: it was by definition based on the definition of a 'proper' distinct from everything else; but now that "proper" has become the whole. It could be that, little by little, it will exhaust its capacity to transform itself and constitute only the space (just as totalitarian as the cosmos of ancient times) in which a cybernetic society will arise, the scene of the Brownian movements of invisible and innumerable tactics. (de Certeau 1984, 41)

Here, in what at first glance looks like an anticipation of the World Wide Web, de Certeau also paints himself into a corner, where the totalizing power he witnessed in globalized New York and postcolonial Europe merged in his mind with cybernetics and network theory, becoming a rather totalizing thought. In addition, the dualism of the borrowed war rhetoric between the strong and the weak among men drafted to war and men who lead them there reaches its end when applied to the rest of the civilian population. What has remained influential, however, is the evoked energy of everyday resistances and hopes that such "small gestures" could become part of something larger to come. This is the hope that has been expressed above by Critical Art Ensemble and Wilding, with the vision of swelling tactics against authoritarian tendencies. Or the hope that we can DIY ourselves into a better world as technology and science become less expensive (Garcia and Lovink's hope above). The key is to refuse to do nothing, because doing nothing, one supposes, is being complacent and, according to de Certeau's war logic, "a loser" in this game. And who wants to be a loser?

The critics of tactical media have raised several important issues. First, even its proponents have pointed out that "minor gestures" of resistance remain, well, minor (Garcia and Lovink mentioned it in the above quotation, even as they proposed tactical media). There is no guarantee that tactics would amount to anything but patting ourselves on our "subversive" backs (a similar argument has been brought up against social practice art or relational aesthetics, too). Second, that even if these gestures are minor, they are still subversive and disruptive, and present a risk to those who create them and implement them, since power operates through force and squashes any resistance as soon as it sees it. This means that only those who are privileged enough or careless enough, or both, can practice such tactical media, because they can use their privilege to get out of danger, and that enables them to claim respect and approval that comes with such "risqué" status. They can afford to be tactical because of their own institutional or other types of "strong" positionality. Third, there are those who disagree with what they see as such a "neoliberal" and defeatist approach to power, a kind of polite minor inconvenience, at a time when we need to organize and work together to dismantle global capitalism. For them, we need to unite to enact our own strong strategies. This kind of critique is often repeated against feminist, LGBTQIA+, and civil rights movements that are labelled as "identity"-based and not comprehending the need to drop their second- and third-level-relevance demands and unite behind the top-level demands of those who critique them.

One of the best and nuanced critiques of tactical media, including the three ones we outlined in an abbreviated form here, is collected by Gene Ray and Gregory Scholette in a special issue of the journal *Third Text* called "Withering Tactical Media" (Ray and Scholette 2008). Their own Introduction to the issue and various contributions are powerful in their documentation and analysis of the strengths of the disruptive potential of the "tactical," but also, in their thoughtful critiques of the limitations of its aesthetic of humor, satire, and prank-type performances of "semiotic" shocks or even the sustained DIY and other tactical projects.

We, however, come to critique de Certeau from a different perspective. First, women constitute almost half of the world population, but they remain invisible in de Certeau's visions of the tactics. Even when he refers to "wisdom," his examples do not see women. Compare that to subRosa's vision of "traditional knowledges" that acknowledges various tactics employed by women as workers:

> A time-honored tactic of workers' resistance has been the withdrawal, or refusal of their labor. For centuries, women, too, have practiced the tactical withdrawal of their sexual services (Lysistrata) and reproductive labor (childfree women, nuns, female mystics) in order to escape patriarchal control. Witches and holy women have used tactics of practicing magic, incantation, and spiritualism as much to escape from allegiance to an unjust system of the gendered division of labor as a means to create an independent living and supply needed services to their communities. (subRosa 2008, 224)

Moreover, as the power of the scientific profession marches forward, traditional knowledges get erased and subsumed, as subRosa illuminated in their *SmartMom* project:

> Without quite realizing the full effects of the magnitude and speed of these cultural changes, much "traditional wisdom" has already been consigned to the historical dung heap and replaced by a survival-driven, first-world reverence for 'smart' technologies; these are proving profoundly disruptive for many populations of different economic, racial, and cultural backgrounds.... The problem with smart technologies is that they circumvent the experience, knowledge, and decision-making power of the user. (subRosa 2008, 274)

Second, de Certeau, in not seeing women, does not see the other institutions that produce and maintain power on the side of the "strong"—family and religion—just as he does not include any life-making and life-sustaining strategies and tactics in his book. Not analyzing these most ancient patriarchal operations, which are closely aligned with governmentality and population management, weakens his own argument, as well as that of those who are trying to create coalitions against authoritarian tendencies of the powerful based on the definitions of tactical media cited above. It becomes unclear, then, who would be included in the unified global opposition to "authoritarian tendencies" and how it is supposed to succeed if women, sexual, crip, and racial minorities' tactics are not mentioned in the definition of the "aggrieved and excluded" or "the weak."

Third, de Certeau's separation of the "strong" and the "weak" does not account for the complexity of "weak" positions and the processes that operate in their disempowerment. The "weak" have become "weak" through the process of *weakening*. It is about taking power away from them. Therefore, we prefer to rename the weak of de Certeau as the "weakened," to account for all the histories of disempowerment as a continuing process. And the strong are strong because they have been *strengthened*. This is what explains the current status quo between the *weakened* and the *strengthened*. This renaming enables us to detach the action of disempowerment from those whom it targets, so that the oppressed are not defined by their oppression and only tactics are available to them. Throughout the book, we seek to show how such a refusal to be defined by seemingly the only option or only the tactic brings forth new pathways for magical lives and strategies of opting out of what de Certeau calls a "game."

Tactical Biopolitics

We can see how cyberfeminists and subRosa redefined tactical media when employing the same aesthetic strategies as other bioartists and tactical media artists of the 1990s–2000s—humorous, anti-establishment, and anti-authority impersonations of technoscience culture; educational and speculative wetware performances; empowerment of the citizens with more affordable technoscience and media, and the rise of the citizen-artist-scientist-technologist-activist model.

In this regard, da Costa and Philip's book (2008), aptly called *Tactical Biopolitics: Art, Activism, and Technoscience*, became an important milestone, a testimony to over a decade of work that brought together the bioart and tactical media communities. This book is also important because it positioned life sciences at the heart of questions of power, capitalism, and activism—the topics that are central to subRosa's work, too. In their *Introduction* da Costa and Philip define tactical biopolitics as indebted to Foucault's analysis of power and also to more recent developments in tactical media and bioart:

> a creative terminological misappropriation, drawing its inspiration from, but not mapping directly onto, two formations: the assembly of resistant cultural practices referred to as Tactical Media, and the intellectual ferment around the history of biopolitics.... In addition to recalling technology practitioners, artists, activists, and theorists, we now call for the inclusion and cooperation of the scientific community. Such an intellectual/political/artistic/technoscientific community is a potentially resistant formation in the heart of postmodern transnational technospheres. The corresponding politics and activities can be understood, however, not only in their most recent contemporary contexts, but also in a much longer historical framework. We owe our understanding of this context to numerous scholars and

> intellectual developments, many of them drawing on the terms "biopolitics" and "biopower." (da Costa and Philip 2008, XVII–XVIII)

Representing this inclusion and cooperation of the scientific community, da Costa and Philip's *Tactical Biopolitics* included an interview with Richard Lewontin, a Professor in Zoology and Biology at Harvard University known for his work in evolutionary biology and genetics. Here, he represents a science-based public intellectual who influenced discussions on the ethical and socio-political implications of genetics, evolutionary theory, and science in general. His critique of genetic determinism as connected to eugenics provides a conceptual framing for our further discussions in the book that center feminist technoscientific approach to reproductive and computational bioengineering. His analysis also highlights the scientific reach of biopower and biopolitics:

> So what I'm trying to say is that if you're a geneticist, you're in the business of studying genes, not phenotype, and the trouble is, until there was DNA sequencing, the only way to study genes was to look at the organism, or maybe its proteins. Then along came proteins, and I spent a certain part of my life looking at proteins. But for most of genetics, between the beginning of modern genetics early in the twentieth century—1910 or so—until 1970, geneticists studied genes by studying organisms. So they have a strong commitment to the view that the organism is made by its genes. Because if you don't believe that, then how can you study the genes by looking at the organism? And they narrowed their investigations down to those cases where there was no ambiguity. … So geneticists say that genes are self-replicating. Genes are not self-replicating. DNA can't do anything. The cell makes new DNA by copying old ones. They say genes make proteins. No, genes don't make proteins. The cell makes protein out of amino acid, using information in the genes. But the genes don't make anything. But that constant reiteration of genes are self-reproducing, they make the organisms—that's what geneticists have always talked about. The social consequence of that has been from the early days that almost all geneticists were strong racists and believed that every aspect of human behavior was caused by genetic difference. (Lewontin, quoted in D'Arcangelis, da Costa, and Philip, 2008, 10)

subRosa cited Lewontin in their own works and supported his anti-deterministic and anti-essentialist positions in relation to the DNA, also critiquing and showing its eugenic histories. Throughout the book, we show how cyberfeminists' use of a tactical biopolitical framework benefited from feminist analysis in its connection to the body and its critique of extractive logics of biomedical research.

Chapter Outlines

In Chap. 1: *The Introduction: subRosa's Legacies*, we outline the major themes of the book and motivations for writing it. We also engage with key aesthetics of subRosa's practice and contributions to Cyberfeminism, Tactical Media, and Tactical Biopolitics.

In Chap. 2: *Terms and Conditions*, subRosa founding member Hyla Willis first sheds light on the influence of typography and media studies on some of the texts accompanying subRosa projects. After that, in the spirit of subRosa's art practice, Willis shares her ongoing engagement with the Terms and Conditions encountered in everyday life that has led her to a form of durational art practice that teases the [im]possibilities for "opting out" of an economy that recklessly extracts consumer data while undermining our leverage to fight its most exploitative practices. She asks if Marshall McLuhan's ideas about media, typography, and the global village might be helpful in understanding the Age of Agentive AI, despite the many shortcomings of early media studies.

Chap. 3: *Domain Errors! A Virtual Roundtable* engages with the twenty-year anniversary of *Domain Errors! Cyberfeminist Practices: a SubRosa project* (Autonomedia, 2002/2003, edited by María Fernández, Faith Wilding, and Michelle M. Wright), that addressed racial, economic, and gendered disparities in biotechnology and the cultural, including early cyberfeminist, theory. A close reading of *Domain Errors!* demonstrates the importance of cyberfeminist conversations to strategies of resistance against the capitalist takeover of life, and how many problems of the coming decades cyberfeminism anticipated. Responding to Aristarkhova's and Willis' questions to the original *Domain Errors!* authors and subRosa members more than twenty years after its publication, María Fernández, Laleh Mehran, Michelle M. Wright, Susanna Paasonen, Amelia Jones, Nell Tenhaaf, and Tania Kupczak offer reflective, generous, and inspiring responses to the tactics proposed, their evolving relationship to cyberfeminism, and update us on their current scholarly, pedagogical, activist, and creative interests.

In Chap. 4: *Magical Life*, Nathanael Elias Mengist and Irina Aristarkhova ask, "Whose/what life is magical?" As a way of answering this question, they offer their visions of creative and just reproductive futures inspired by subRosa and Wu Tsang's non-binary worldbuilding of "undoing gender," alchemic history in the present, postcolonial technoscience studies (Rajani Sudan), Black Feminist Thought (Dorothy Roberts, Zakiyyah Iman Jackson, Tiffany Lethabo King), reproductive mutual aid of Black kinship (Cynthia R. Greenlee) and Argentinian abortion joy (Lea Happ). In the Chapter, Mengist and Aristarkhova center two of subRosa'sworks—*Yes Species*, and *Sex and Gender in the Biotech Century*—within the contemporary framework of feminist technoscience and biopolitics of knowledge production. They also offer their own creative contributions and a method they call *compassionate inquiry* as they connect subRosa's legacy to concerns over "life-giving," "population-forming," and "generative" bodies, with ARTs and industrial/business processes (e.g., "decision trees") in the quest for magical life.

In Chap. 5: *Techno Rant of the Cyborg Elder*, a founding subRosa member, Faith Wilding, revisits her "Rant of the Menopausal Cyborg"—written a quarter century earlier for *Domain Errors!* -- by offering a new poetic and evocative contribution. Wilding claims the persona of a *Cyborg Elder* to trace the changes in her experience of and attitude towards ageing, ageism, memory, body, and technology. Living through the consequences of the predominantly ableist nature of technoscience, Wilding goes back to the technologies of her childhood in Paraguay, such as the use of radio or plant parts as technology, and connects their embodied memory to various forms of writing systems and computation, their role in remembering, and challenges with producing a community in her present. Wilding especially focuses on the tension between the elemental and the technological.

In Chap. 6: *subRosa Manifestations: A Chronology of Particular Projects, with Notes*, Hyla Willis takes the reader through a tour of some of the group's key works and shares new details and nuanced documentation of how several key projects were initiated, researched, reworked, and refined. Descriptions of key subRosa projects are summarized in chronological order and annotated with descriptions and bibliographic references with students, researchers, artists, and other readers in mind.

Overall, this book seeks to contribute to the fields of art & design, bioart, art, technology and science, bioethics, new media art, art history, performance studies, digital studies, media theory, cultural studies, speculative and science fiction, science and technology studies, feminist technoscience, and feminist, gender, and sexuality studies.

References

Aristarkhova, Irina. 2012. *Hospitality of the Matrix: Philosophy, Biomedicine, and Culture*. Columbia University Press.

Browne, Simone. 2015. *Dark Matters: On the Surveillance of Blackness*. Duke University Press.

Browne, Simone. 2017. *"Get at a way of telling": On Black Net.Art Actions*. https://rhizome.org/editorial/2017/may/11/get-at-a-way-of-telling/

Catts, Oron (ed.). *The Aesthetics of Care? : the artistic, social and scientific implications of the use of biological/medical technologies for artistic purposes*. Presented by SymbioticA, the Art and Science Collaborative Research Laboratory and The Institute of Advanced Studies, University of Western Australia. SymbioticA, University of Western Australia, 2002.

Clynes, Manfred E., and Nathan S. Kline. 1960. Cyborgs and Space. Astronautics, 26–27, 74–76.

Critical Art Ensemble. 2001. *Digital Resistance: Explorations in Tactical Media*. Autonomedia.

Da Costa, Beatriz and Kavita Philip. 2008. "Introduction." In *Tactical Biopolitics: Art, Activism, and Technoscience*, edited by Beatriz da Costa and Kavita Philip. The MIT Press, 2008.

De Certeau, Michel. 1984. *The Practice of Everyday Life*. University of California Press.

Falconer, Rachel. 2013. "The Body Politic of subRosa." *Furtherfield online magazine*, July: n.p.

Fernández, María, Faith Wilding, and Michelle M. Wright, eds. 2002. *Domain Errors! Cyberfeminist Practices*. Autonomedia. Available at http://refugia.net/domainerrors

Fernández, María, Irina Aristarkhova, and Coco Fusco. 2002. "Undercurrents: A Dialogue." *Fine Art Forum* 16 (8). http://web.archive.org/web/20061008031138/www.fineartforum.org/Backissues/Vol_16/faf_v16_n08/reviews/feature.html

Haraway, Donna. 1991. *Simians, Cyborgs and Women: The Reinvention of Nature*. Routledge.

Hung, Christina N. 2005. *All Things Grow with Love*. https://www.instagram.com/p/Bt4aO5pgU8O/

Johnson, Jessica Marie. 2018. "Markup Bodies: Black [Life] Studies and Slavery [Death] Studies at the Digital Crossroads." *Social Text* 137, 36 (4): 57–79.

Jordan-Young, Rebecca. 2010/2011. "The subRosa Collective: Cyberfeminist Interventions." *S&F Online*.

Marchevska, Elena & Valerie, Walkerdine. 2016. "Editorial." *Studies in the Maternal* 8 (2): 10. doi: https://doi.org/10.16995/sim.236

Motter, Jennifer L. 2011. "Feminist Virtual World Activism: 16 Days of Activism Against Gender Violence Campaign, Guerrilla Girls BoardBand, and subRosa." *Visual Culture & Gender* Vol 6.

Ptqk, Maria. 2017. "TECNOCIENCIA, FEMINISMOS Y BIOPOLÍTICA TÁCTICA. Contextos y prácticas del colectivo subRosa." UPV-EHU Universidad del País Vasco, Facultad de Bellas Artes / Departamento de Arte y Tecnología – Arte Ederren Fakultatea / Arte eta Teknologia Saila. Tesis doctoral. https://www.mariaptqk.net/tecnociencia-feminismos-y-biopolitica-tactica-contextos-y-practicas-del-colectivo-subrosa/

Ray, Gene and Gregory Sholette. 2008. "Introduction: Whither Tactical Media." *Third Text*, 22 (5): 549–558.

Russell, Legacy. 2020. *Glitch Feminism: A Manifesto.* Verso.

Seu, Mindy. 2023. *Cyberfeminism Index.* Inventory Press.

Stocker, Gerfried, ed. 2000. *NEXT SEX: Sex in the Age of its Procreative Superfluousness. Sex im Zeitalter seiner reproduktionstechnischen Überflüssigkeit* (Ars Electronica). Springer.

subRosa. 1998. *Manifesto.* http://cyberfeminism.net/about/manifesto/

subRosa, Irina Aristarkhova, Margaret Tan, and Adeline Kueh. 2003. "Tactical Media Lab Report to N5M4 Editorial Board on Matrixial Technologies, Singapore, January 1–16, 2003." February 18. http://www.tacticalmediafiles.net/n5m4_journal/indexda63.html

subRosa. 2008. "Common Knowledge and Political Love." In *Tactical Biopolitics: Art, Activism, and Technoscience*, edited by Beatriz da Costa and Kavita Philip. The MIT Press.

Timeto, Federica. 2010. "Unmasking the Theatre of Technoscience: The Cyberfeminist Performances of subRosa." *Feminist Media Studies* 10 (2): 244–248.

Wilding, Faith. 2003. "Rant of the Menopausal Cyborg." In *Domain Errors! Cyberfeminist Practices,* edited by María Fernández, Faith Wilding, and Michelle M. Wright. Autonomedia. http://refugia.net/domainerrors/DE3i_rant.pdf

Wilding, Faith and Hyla Willis. 2016. "SmartMom Rebooted: A Cyberfeminist Art Collective Reflects on its Earliest Work of Internet Art." *Studies in the Maternal* 8(2): 17. Special Issue "Performing Everyday Material Practice," edited by Elena Marchevska and Valerie Walkerdine. doi: https://doi.org/10.16995/sim.229

Yamin, Johann and Samantha Yap. 2023. "Mindspace Presentation #17, in Conversation with Margaret Tan." *Hothouse: Media Art in Singapore.* https://h0t.house/presentations/margaret-tan

Yap, Samantha and Teow Yue Han. 2021. "Mindspace Presentation #6, in Conversation with Irina Aristarkhova." *Hothouse: Media Art in Singapore.* https://h0t.house/presentations/irina-aristarkhova

CHAPTER 2

Terms and Conditions

Hyla Willis

They came to harvest our pain.

– Barbara Kingsolver (2025)

This essay will look at the influence of typography and rhetoric on certain of the art collective subRosa's projects and share my own *durational practice*—informed by both typography and performance art—in which I confront *Terms and Conditions* encountered with increasing frequency in my everyday life. In the spirit of subRosa's cyberfeminist inquiry, I hope to tease out possibilities for understanding and resisting exploitative new conditions, to find utility in what presents itself as futility. I will show how typography is utilized within commercial products and services to facilitate "points of extraction" from "users," and argue that there are lessons to be taken from Marshall McLuhan's analysis of movable type that may be useful in recognizing the spread of legal traps hidden in many apps and in grappling with emerging forms of artificial intelligence.

subRosa's earliest projects included net art, pamphlets, and interactive soft sculpture, but soon evolved into complex, media-rich, interactive performance art, often customized with research connecting it to the specific sites and situations where the work was presented (see Chap. 6: *subRosa Manifestations* for a descriptive chronology of projects). The collective maintained a prolonged interest in biotechnology and the fertility industry as sites of impact on the bodies and labor conditions of women,

H. Willis et al., *Night Sweats: Cyberfeminist Practices*, Palgrave BioArt,
https://doi.org/10.1007/978-3-032-06955-9_2

particularly, and as a fruitful ground for critical, experimental feminist art, bringing these interests to discourses of cyberfeminism and bioart.

Many of subRosa's projects contained a multitude of other *micro-projects* that were revised, reused, and remixed with regularity—some of which could stand on their own as discrete works of art worth a closer look. These *projects-within-projects* took the form of written texts, broadsides, websites, "swag," and more. All were created and distributed with an awareness of how the various histories of media, biotech, and surveillance technologies informed the visual and written forms they took.

Although subRosa members would convene in Pittsburgh for in-person brainstorming sessions on new projects, the work for most shows would come together in a patchwork fashion. Not long after its founding, several members had to relocate for jobs, which meant that various project components would be developed in different geographic locations. Members would check in with each other via email or work in pairs to draft a text, outline a budget, sketch a costume design, build a prop, or create graphic designs.

Here I will pause to acknowledge a strange feeling of whiplash as I combed through my archives for this book. In early 2025—against the backdrop of Donald Trump and Elon Musk's decimation of US-backed projects that ease hunger and fight disease around the world—I am digging through conference, festival, and gallery materials and grant applications from the past 25+ years. This means I'm also revisiting the sometimes-fraught histories of domestic and international arts programming. Government- and NGO-funded humanitarian and cultural exchange initiatives are almost always driven by complex agendas and changing political winds. We artists, alongside our hosts, the funders, and our intended audiences, should always strive to grapple with the uncomfortable complexities of "soft power," yet still forge ahead as artists who make art. While doing the work of cultural production—soft power's only real asset—we should critique and address the politics of our funding streams beyond merely entertaining shame and anxiety about which keywords and images to include or concepts to mask.

Each member of subRosa brought her unique prior skills and knowledge to the group. Many of us were already seasoned teachers, artists, and media makers. I had specialized experience as a typesetter and graphic designer working in the publishing industry, who studied journalism and video art before completing an MFA. In this essay, I will draw on my expertise in typography and visual rhetoric to unpack some of subRosa's micro-projects and introduce and frame my ongoing durational practice of "opting out."

Am I Typographic Man?

In 1962, Canadian philosopher Marshall McLuhan wrote *The Gutenberg Galaxy: The Making of Typographic Man*, which popularized the contested term *global village*. In it, he examines how the invention of movable type—which you might think of as the *automation* or *industrialization* of writing systems that use the Latin alphabet—altered Western consciousness by expanding the stewardship and distribution of knowledge (McLuhan 1962). The movable type printing system allowed for sets of individual letters to be "locked up" into page layouts for replication on a printing press, then be disassembled and the components reused for future projects. This was faster, cheaper, and required different skills and materials than the hand-copying of manuscripts by scribes or the meticulous artisanal carving and engraving of printing plates.

Three hundred years ahead of the British Industrial Revolution, Gutenberg's developments (which also included innovations in paper, ink, metallurgy, and the printing press) gave rise to the mass production of books in Latin and changed the means and speed by which information could be compiled, retrieved, and disseminated.

McLuhan viewed the shift from oral and auditory traditions, to handwriting, to books printed with "movable type," as causing profound changes to human cognition and social organization at each turn. Decades ahead of the World Wide Web, he predicted the *global village*—a convergence of manuscript traditions and "electronic media"—that would lead back to a form of social interconnectivity having more in common with earlier oral and tribal traditions than with the individualism of print culture and its emphasis on the visual. In *Understanding Media,* he stresses the importance of studying the nature of media itself as much as the content it carries, showing his continued insight into how technology as an extension of man's senses is always already reshaping cognition and society (McLuhan 1964). McLuhan is playful, cross-disciplinary, and presents a literary mosaic that views typography as a centralizing tool that helped formulate modern concepts like individualism, nationalism, and capitalism. At times, his *typographic man* is too universalizing of a concept, and the *global village,* while less universalizing, underestimates the sticky nature of typography and how it works in tandem with extractive capitalism to subsume the interconnectivity generated so freely by the villagers. Sarah Sharma and Rianka Singh redress some of McLuhan's shortcomings in an edited volume, *Re-Understanding Media: Feminist Extensions of Marshall McLuhan* (Sharma and Singh 2022), and in Sharma's programming of the McLuhan Centre for Culture and Technology during her leadership there. In Sharma's words in the book's introduction, there is "potential in understanding that the technological is a specific vector of power that demands a feminist understanding" (Sharma and Singh 2022).

Yet McLuhan's ideas about the automation of the [Western] writing system and its impact on literacy and numeracy—as well as his nontraditional ideas about what constitutes media—can lend us insight when considering the current wave of automation: the rise of generative and agentic AI that can "write" and "read" [i.e., summarize] texts, and that can emulate boilerplate conversations of the sort that are endemic to "customer service" interactions, based on machine learning of consumer psychology.

The designs of Johannes Gutenberg's first fonts for the printing press were rooted in blackletter calligraphy, in visual form and in scale. This detail is worth noting because they all but replaced the work of scribes who had been replicating manuscripts by hand. Gutenberg's inventions enabled the mass reproduction of texts, and the typography soon broke free from the blackletter tradition, giving rise to typeface designs that are still in use today. "Adobe Jenson" is one example—a digital font created by Robert Slimbach in 1996—that is a revival of a design first cut in metal in 1470 by Nicolas Jenson, but that is still utilized in contemporary graphic design (*Adobe Jenson*, in "Adobe Fonts," https://fonts.adobe.com/fonts/adobe-jenson).

Donald J Trump's team used a typeface designed in the 1890s—Akzidenz-Grotesk—for his 2016 and 2020 US Presidential campaigns (Monotype 2019), switching to Gotham—designed in 2000 for *GQ Magazine*—for the 2024 campaign. To the untrained eye, these are nearly equivalent blocky sans-serif headline fonts, but Gotham telegraphs a subtle contemporariness you might recognize from its widespread use in Barack Obama's 2008 Campaign, in addition to holding hidden New York City visual references (Schwarz 2022). If you feel stumped differentiating between the two examples shown below, look for clues in the counter forms of the R and P, or in the center vertex of the M (Fig. 2.1).

What is the point of including such a trivial-seeming comparison between two typefaces here? It is because typography merits multiple dimensions of analysis. In the preface to *Gutenberg's Galaxy*, McLuhan turns briefly to Claude Bernard, a nineteenth-century practitioner of vivisection, to describe the difference in the sciences between *observation*—"noting phenomena without disturbing them"—and *experimentation*, which involves imposing a variation or disturbance on the phenomena (McLuhan 1962, 3). I have leaned into the former to help you observe the subtle design differences between two "engineered" typefaces. Notice that the individual glyphs are carefully designed characters and not mere imitations of calligraphic forms drawn with chiseled pen nibs held at precise angles, as seen in early blackletter fonts.

In my choice of visual samples, I also hope to invite a closer reading, a sort of literary analysis: Why the switch in campaign typefaces? Why the Trump logos in this chapter? Why now?

Returning to Gutenberg: thanks to movable type, book publishing saw explosive growth in the West. By the sixteenth century, the typefaces, sizing, and visual alignment that feel familiar to the present day, in the dozens of languages that use the Latin alphabet, became expected conventions. These "norms" have been so continuously baked into visual communications that it is now common to experience "authority bias" in response to the typographic forms we encounter in everyday life. Before reading the texts set before us, it is common to make judgments about the written content based on its typesetting alone.

When using satire for tactical media projects, the mimicry or hijacking of typographic conventions is one place artists turn to when setting up the ruse. These norms—known as *typographic scale* (Bringhurst 1996, 42)—are not hard for the layperson to tap into: the list of pre-ordained typeface sizes you can still find in popular word processing software (8pt, 10pt, 12pt, 24pt, 72pt, etc.) was standardized centuries ago by those European printers working with limited sets of individual cast metal letters designed for reuse in future books and broadsides. The printing trade at that time

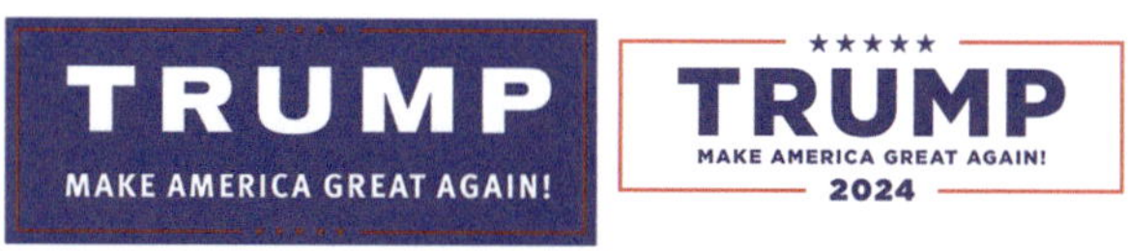

Fig. 2.1 A comparison of two typefaces used for the Trump US presidential campaign graphics. The left image features Akzidenz-Grotesk (Berthold circa 1890) and the right features Gotham (Hoefler & Co. 2000).

could not access the infinite scaling and variety possible with today's desktop publishing. They were limited to wrangling bulky wooden cases containing cast metal letterforms: one 50–70lb drawerful for each point size, in every typeface.

Typography is everywhere: for half of a century, broadcast news telegraphed its authority in part through the typographic and rhetorical conventions of the *Chyron*—that scrolling ticker on the lower third of the screen (NBCU Academy 2024). Today, screen-based typesetting and most forms of signage—although they have been "liberated" by technology from 500 years of rigidly-scaled movable type—still lean into entrenched reader expectations about what constitutes a "main text," and what words might be comfortable (if not wise) to skim or outright ignore. This is true whether viewing a website on your computer, an app on your handheld phone, or checkpoint signage at an airport or border crossing.

The Terms and Conditions, Privacy Policies, and Disclaimers that come with most operating systems, software, and apps are known as *Standard Form* (i.e., boilerplate) *Contracts*, and they apply to a mind-boggling number of global transactions daily. Recall for a moment the last time you installed or updated an app on your smartphone. To use the app, you must tick a button or box indicating that you have read and agree to the *Terms of Service* and the *Privacy Notice*. Your agreement to this legal arrangement is easily enforced by blocking your use of the app or website until the terms have been accepted. As an adult with agency, you can take it or leave it, right? No one is forcing you into these voluntary arrangements, or so it would seem.

Take, as an example, a banking app you've voluntarily installed on your smartphone. On first launch, you are prompted to create an account and click that you have read and agree to whatever terms are presented. They're presented again any time there is an update to the app. Apps that provide powerful access to your financial resources often have a security measure built in: you cannot use older versions of the app beyond a certain point to protect your data from malicious actors. New versions of these apps roll out at odd times, often as "patches" in response to major security breaches. It is easy enough to scroll through these agreements and tick a box indicating that you have read them, even if you have not.

In this moment of escalating deportation raids in the US, I recently purchased tickets to a music festival benefitting the immigrant community in my city. The online transaction appeared straightforward, and the UI design made it easy to bypass the *Terms of Service* (29 pages containing

1600 words) and *Privacy Notice* (20 pages containing 8280 words), but still complete the purchase. My clicks constituted my consent that "information about yourself and others" would be harvested, to include details about the beneficiary, fundraiser goals, and "situations or other circumstances that motivate a fundraising request."

There is no singular answer as to whether or how Standard Form Contracts should be "kept honest"—in their written content or visual form—or how much real-time parsing of legal terms is feasible for the harried consumer. In addition to the terms, users must grapple with the matter of privacy. Tracking pixels are embedded in files and act like spies that send reports back to their host about your location, devices, and interactions. Cookies are stored in your browser and track what you do there. These encroachments on privacy are handled very differently in the EU, the US, and China. Businesses, academic researchers, and governments have differing views on how markets might regulate themselves in this regard. In the European Union, the General Data Protection Directive (GDPR) safeguards privacy as a fundamental right. In the US, the burden of assessing risk is largely shifted to the individual consumer to sort out. It is confusing, time-consuming, and sometimes humiliating to make informed decisions about which losses of privacy and legal rights are worth the risks for gains in convenience at each transaction involving data surveillance. Furthermore, what is at stake pans out differently, depending on your personal legal status and leverage.

A related question of greater consequence: What is the equivalent of a Standard Form Contract for individuals who are crossing borders? What are reasonable expectations for tracking the changes to one's own status (after all, not everyone possesses citizenship), or the status of a border (ports, free trade zones, outer airspace, and occupied territories are especially complex), and who becomes the steward of the information gathered at these real and virtual checkpoints? What happens to a country's data when its assets are overtaken?

Technology providers routinely seek our voluntary consent to forfeit certain of our rights so they can shield themselves from litigation, yet in the age of the Internet of Things, the sheer volume and complexity of these consequential texts—and their concealment through diminutive typography—urges our capitulation. "The fine print" is not insignificant to our lives—it typically contains alarming and disempowering stipulations that are a precondition of engaging with products and services. In the United States, it is commonplace to agree to a daily multitude of terms and conditions that one cannot be bothered to read or listen to with the full attention they deserve. A string of recent lawsuits and position papers sheds light on this growing problem, and I will discuss a few of them below.

UX and UI designers use the term "pain points" to categorically identify avoidance behavior inspired by bad or overwhelming interface design (Gibbons 2021). With the goal of preventing users from abandoning online transactions out of frustration, a benevolent design approach might guide the user through an annoying but necessary series of actions. Unfortunately, these pain points can also be a window for corporate or state opportunism. In late 2022, the US-based Federal Trade Commission released a report warning about "dark patterns" used by companies to "trick or manipulate consumers into buying products or services or giving up their privacy." (Federal Trade Commission 2022). Something as simple-seeming as buying event tickets with your phone can unleash unseen tracking of your every move in real and virtual space. A related dark pattern—making subscriptions extremely difficult to cancel—has earned the nickname "sludge." The satirical film *Mountainhead* mocks this business model via mention of a mindfulness app whose profits are derived from the difficulty of cancelling the service (*Mountainhead* 2025).

One recent summer day, I encountered a very dark pattern inside a Pittsburgh parking garage. I pulled into the garage with ease for a 15-minute errand at a nearby shop, and was surprised at the lack of the usual machine offering a time stamped ticket. On exit, I discovered that the only available means of payment was through an app I would be required to load on my phone, whose *Terms of Service* would strip me of legal rights to hold the firm accountable for any civil or criminal act within the garage other than through individual arbitration arranged by the company. The Privacy Notice explained that I would be opted in to targeted marketing on my phone from all the businesses in the nearby development. A sign at the garage exit warned that my car plates had been recorded [without notice] on entry and that any attempt to evade payment would be met with a large fine. I was trapped! I exited the garage without paying and phoned the garage company from a nearby loading zone. *"Please, I don't agree to the terms of your app. How can I pay without putting your app on my phone? There was no warning of this at the garage entrance!"* After working my way through several layers of management, I was told there would be no possible means of payment or avoiding the penalty other than by loading the app, but that I could immediately delete it when I was done with the transaction. This anecdote contains many clues about me: the fact that I have a car and a smartphone, do my own errands, and have the audacity and the time to push back when I feel wronged. I would like to persuade you that it is also an example of durational performance art, and a means of reclaiming what companies patronizingly call the *user journey*, but which I think of as life itself.

Acknowledging the affective and abusive nature of Terms and Conditions by making the type small—or altogether absent—eases the "pain points" and encourages consumers to ignore important texts while protecting companies from running afoul of the law. The histories that prompted companies, governments, and citizens to develop the practice of using Terms and Conditions and the current legal landscape of their regulations are highly variable and dependent on the prevailing laws where one is located. In the US and some other jurisdictions, it is furthermore reflective of companies' desire to preemptively shield themselves from lawsuits. *Observing* and doing a *close reading* of the typography cannot entirely protect you from the abuse, but the experiment of casting such actions as a *durational practice* is a form of creative refusal. For starters, you will elevate your own status from that of a *consumer* to that of a *reader*.

A closer look at subRosa's performances and publications reveals the frequent use of humorous and fictional *fine print* and other rhetorical conventions of typography and advertising, such as parallel texts, annotations, glossaries, and marketing slogans. The group wanted to connect the dots, together with the audience, linking ideas about Empire to the American Eugenics movement, and to the hijacking of feminist rhetoric about bodily choice by new forms of consumerism. subRosa brought emphasis on ARTs to its cyberfeminist art practice. The marketing and legal language surrounding the fertility industry in the US has been fair game for critique, and subRosa satirized both while looking critically at reproductive justice and inequity in the sciences by examining both the words and their visual presentation. Thus, subRosa's satirical spins on waivers, release forms, warnings, and other boilerplate texts—on closer read— aimed to shed light on the incursions into civil and legal rights presented by the texts they aimed to mock.

In one example, subRosa's *Cultures of Eugenics* pamphlet (Fig. 2.2) created for the *U-Gen-A-Chix* performance (subRosa 2003, see Chap. 6) opens with an excerpt from the "forward-looking statement" found on the website of the biopharmaceutical company Viragen. Forward-looking statements are an aspect of US business law requiring transparency to potential investors—to distinguish facts from conjecture, and to communicate risk. In essence, they are a type of "fine print" that quietly warns investors that marketing language may discuss or predict business developments that might not be achievable. In the early 2000s, Viragen was developing a flock of Transgenic Hens that would lay "virtually unlimited" numbers of eggs containing therapeutic proteins and the company was courting investors online.

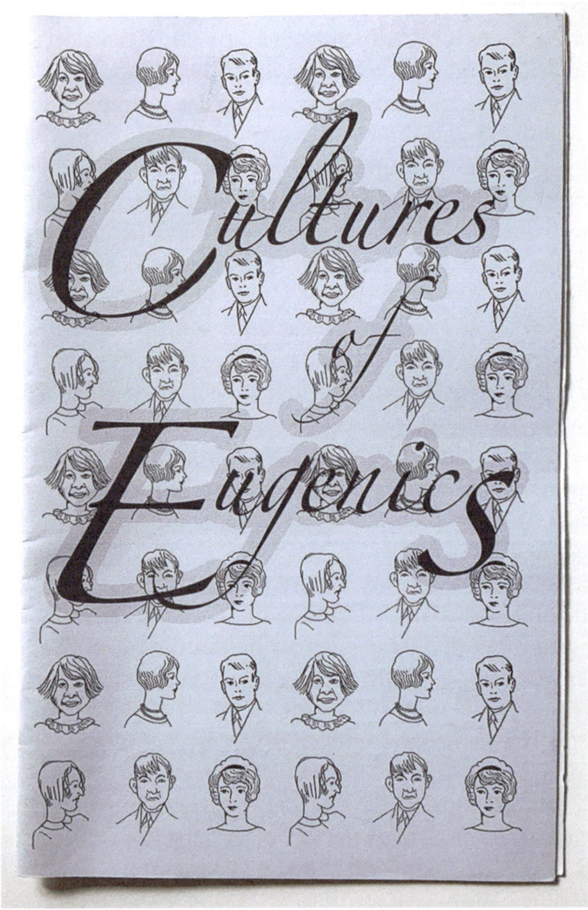

Fig. 2.2 The cover of the *Cultures of Eugenics* pamphlet featuring figures from a Stanford-Binet test. 16pg folded half-letter booklet (subRosa 2003).

> *Disclaimer: This website may contain information that includes or is based upon forward-looking statements with the meaning of the Securities Litigations Reform Act (SEC) of 1995. …You can identify these statements by the fact that they do not relate strictly to historical or current facts. They use words such as "anticipate," "estimate," "expect," "project," "intend," "plan," "believe," and other words and terms of similar meaning.* (Excerpt, Viragen, quoted in subRosa 2003)

subRosa closes the pamphlet with a parody text:

> *Backward-looking statements give an overview of historical realities, through the filter of the interests, ideology, access to information, and power of the person collecting and re-presenting the information.… they are carefully documented and seem to come from legitimate resources that may or may not have a direct interest in the argument being advanced.*
>
> *Any or all of our backward-looking statements here or in other publications may turn out to be wrong. They can be affected by inaccurate assumptions or by known or unknown risks and uncertainties, such as the disappearance of witnesses, the burning of libraries, and systems of oppression that force illiteracy upon generations. Many such factors will be important in determining our actual recorded histories.* (subRosa 2003)

By 2007, Viragen was bankrupt. The company ultimately abandoned their work developing transgenic therapeutic egg whites to focus on anti-cancer drugs such as Multiferon (Fierce Biotech 2007).

In 2003, subRosa updated the "Flesh Worth Form" first developed for the *US Grade AAA Premium Eggs* and *U-Gen-A-Chix* performances. For its use in *International Markets of Flesh*, a redesign aimed to make the form seem more official, bureaucratic, and authoritative by imitating the typography and layout of the *1040 EZ* IRS form used by most tax-paying workers in the US from 1982–2018 (Fig. 2.3).

Marshall McLuhan frames Typographic Man as a visual creature and linear thinker in *Gutenberg's Galaxy*. While graphic designers might easily manipulate our presumptions about content through clever typography, our reliance on "reading" in the conventional sense is becoming overwhelmed by the bombardment of commands on our attention. Cleverly, corporations have moved on, capitalizing on their presumption that we have given up on reading and will just "trust the typography." The fine print has become a Trojan Horse, enabling the Standard Form Contract to morph into something that quietly strips

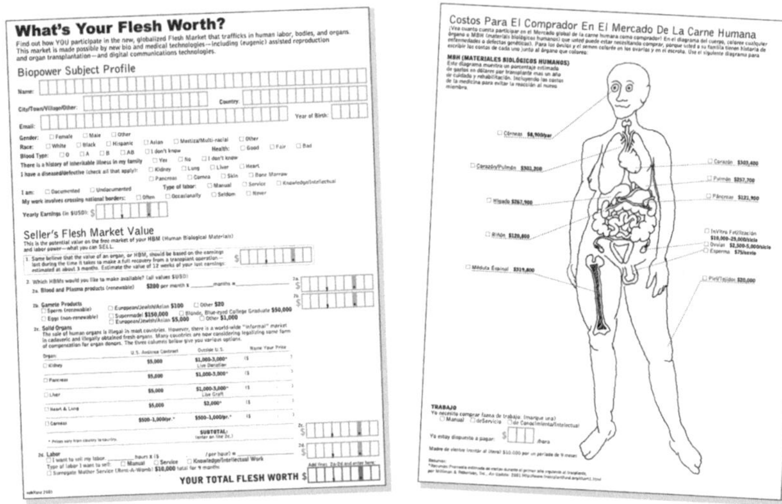

Fig. 2.3 Two sides of the Flesh Worth Form used in several subRosa performances (subRosa 2003).

away your ability to fight the potential mishandling of both your data and your embodied self.

To Whom Do Our Bodies Belong?

The US and New Zealand are the only countries that allow direct-to-consumer pharmaceutical advertisements. The US media is saturated with them (DiStefano et al. 2023). On time-based platforms, the disclaimers are sonically presented. In a TV ad for a prescription drug used to treat diabetes and chronic kidney disease, a cheerful color-coordinated musical dance number unfolds in an office setting, while a voiceover actor warns of side effects including an "increased risk for lower limb loss" (Jardiance Commercial 2024).

Thoughts of suicide and clay-colored stools are among the risks associated with the prescription weight loss drug Wegovy, a cheerful voice explains while a parade of overweight but joyous Americans forms on the streets of a historic Anytown business district (Wegovy Commercial 2024–2025).

The cognitive dissonance transmitted by these ads and many more like them is a staple of daily life in all sectors of American culture penetrated by advertising, which is in most spaces, including the home.

In presentations about its work, subRosa has sometimes teasingly asked audiences, *to whom do our bodies belong?* While we may not draw the same conclusion reflective of the biblical source of this juicy question (1 Cor. 6: 12–20 (NKJV)), subRosa's projects do put this question in front of audiences—in rhetorical form and often in a spirit of collaboration with them—joining theorists Federici, Shildrick, Ehrenreich and others to draw particular attention to the ways that the female reproductive body and feminized labor have long been the subject of extractive economies.

The *SmartMom* website, crafted by subRosa in 1998, was a dystopian speculative idea launched before social media and biometrics existed in their current forms, but which anticipated the extremes to which corporate and state powers would harness the forthcoming "attention economy" (See Chaps. 1 and 6). An early example of net art, subRosa imagined *SmartMom* as a sensate pregnancy dress inspired by actual remote battlefield medical textiles that were being developed at Georgia Tech, a major research university. subRosa's version peddled a fictional "wearable" for gestating women, one that would subjugate the female body to be optimally reproductive for state and corporate overlords, with every physical and emotional output measured, quantified, and leveraged. It was a cynical mashup, imagined several years ahead of the post-9/11 government surveillance programs. And ahead of the lifesaving though hackable (Jaret 2018) biomedical devices like pacemakers and insulin pumps that can be accessed by the wearer's own cell phone, and Neuralink's implantable generalized brain–computer interface that is currently in clinical trials.

In today's world of influencer culture and continual online marketing surveillance, the figure of the *Tradwife* [a portmanteau of "traditional wife"] could at first glance seem like a satirical tactical media art project conjured by second-wave feminists, but it is not. This online trend encourages white Christian married mothers to monetize their 1950s-era traditional housewife lifestyles on popular social media channels such as Instagram and TikTok by posting slice-of-life videos containing commissioned product placement. Notwithstanding the retrograde socioeconomic roles on display, Tradwives are real entrepreneurs—businesswomen in the gig economy. The figure of the Tradwife offers an aspirational spectacle (aspects of which were presaged in the pioneering artworks of Lynn Hershman Leeson) but is in fact a savvy networked mechanism extracting, categorizing, remembering, and selling a record of *your* habits and fractured attention.

The fictional product *SmartMom* encloses the female body to extract and control data. The Tradwife sells the fantasy of reclaiming one's time and family—on the surface, a kind of *choice feminism*—while quietly taxing your readership with surveillance. *Slurp slurp slurp.*

Extractive digital economies "mine" our bodies, our labor, and our attention to build corporate wealth and consolidate power. Like the first imperial extractive powers, big tech makes discoveries and usurps value created by a population, denying it any claim. It then sells this value back to the population while chipping away at legal remedies to any injustice or harm caused. When exploring how copyright and intellectual property might evolve in the digital age, in his 2004 book *Free Culture*, Creative Commons founder Lawrence Lessig gives the example of Disney films that are based on common folk tales, but now fiercely protected by IP claims. Lessig is not anti-property per se but argues that for cultural sustainability, some rights ought not to last forever (Lessig 2004).

In subRosa's *Cell Track: Mapping the Appropriation of Life Materials*, dark red circular graphics overlay a larger-than-life anatomical figure and give examples of successful and failed legal bids to wrest biological intellectual property back from unwanted corporate takeovers (See Fig. 6.16, Chap. 6).

New forms of extraction and privatization are currently being tested by generative artificial intelligence chatbots (for example, ChatGPT, Midjourney, and Grok) that are "trained" on knowledge and creative work done by humans. The companies developing these chatbots are sometimes licensing training data or scraping it from the public domain, and other times just stealing or "borrowing" it without permission.

So, *to whom do our bodies belong?* No one belongs to Ancestry.com, the largest for-profit DNA-testing company in the world, which is owned by The Blackstone Group. Nor do we belong to the online shopping behemoth Amazon, whose subsidiary holds the world's largest share of internet cloud infrastructure, and hosts biometric databases for the US Customs & Border Patrol and thousands of other government agencies. What about our faces? Do they belong to Clearview AI, a company built on data illegally harvested from 30 billion images posted to Facebook and other social platforms to create a privately held biometric database rented and shared globally with law enforcement agencies? What happens when the security of these troves of sensitive data is breached?

Do our faces belong to the venues that capture biometrics as we enter? These include numerous American football stadiums and Madison Square Garden, which have used Clearview AI's database and algorithms to retaliate against unwanted litigation by banning certain law firm employees from entering music concerts and other events (Hill and Kilgannon 2022).

One of Clearview AI's co-founders—American businessman Richard Schwartz (Hill 2020)—built his fortune in the private welfare-to-work sector, following his time gutting New York's welfare system as a public servant under the mayorship of Rudolph Giuliani (McGowan 1999).

Clearview AI has achieved stunning and rapid success by peddling facial recognition for its alleged convenience and safety. They stole two billion images found on the Russian equivalent of Facebook (V Kontakte). They offered Ukraine free training and access to this biometric trove "to uncover Russian assailants, combat misinformation, identify the dead, and vet people at checkpoints" (Dave and Dastin 2022). A side benefit of this "generosity" was the accrual of masses of Ukrainian faces to Clearview AI's private database.

In the words of Albert Fox Cahn of the Surveillance Technology Oversight Project in New York: "once you introduce these systems and the associated databases to a war zone, you have no control over how it will be used and misused" (Dave and Dastin 2022).

We must proactively ask: what happens to the data when Russia's "military operation" in Ukraine concludes?

Opting Out: My User Journey

Over the years of working on *SmartMom*, *Cell Track,* and other subRosa projects, I was personally becoming more sensitive to the "privacy policies" and "terms and conditions" encountered in my everyday life. As I navigated some health crises of my own in the mid-00s, I began altering and annotating the policies presented to me at medical appointments rather than just signing them outright.

Altering contracts before signing them was already a familiar practice to me, having worked in the worlds of graphic design and architecture, where it was not uncommon to scratch out contract clauses to which one did not wish to agree, yet sign in agreement with the remaining terms. After all, a contract is not *the law itself*, but a mutually negotiated, binding agreement enforceable by law. One should not sign contracts to which one does not agree, and it is lawful to step away from signing such agreements.

In fact, some states (and to a lesser degree, the US federal government) offer protections in the form of a "cooling off" period where consumers can cancel high-pressure, high-penalty contracts such as for home repairs or gym memberships, up to three days after they have been signed (see Pennsylvania Attorney General's "Right to Rescind" law, 2025).

I've been tracking and pushing back against the ever-growing constellation of encroachments I notice in my everyday life. This started innocently enough, by closely reading the boilerplate privacy waiver I was required to sign before receiving gynecological care. It gives Pittsburgh's dominant medical system—The University of Pittsburgh Medical Center [UPMC]—the right to claim my biological material as its intellectual property for research. They say the material is disidentified, but this is an empty promise in an age of DNA sequencing and shoddy data management. Furthermore, it minimizes the patient's wish for privacy as if privacy were merely a vanity project.

For years, I would strike through the most current version of their "consent to be treated" form and annotate it in handwriting.

> *If applicable, I give UPMC permission to appropriately dispose of any specimens/tissue (such as blood samples, PAP smears, skin tags, etc.) taken from my body. Once disposed of, these specimens/tissue cannot be retrieved. I understand and agree that UPMC and its designees may use such specimens/tissue as part of its educational activities. I understand that state and federal law allows UPMC to use specimens/tissue for research purposes without my authorization if my identity is not linked to the specimens/tissue.* (UPMC Patient Handout, Willis personal archive 2025)

The statement went on to explain that disidentified specimens used in research would become the intellectual property of UPMC, which is the piece I objected to. In handwriting, I offered my leftovers to any research conducted in the public domain, rejecting the privatization of my tissues. My amendments worked, until they didn't. Eventually, the forms went digital with no possibility for edits. Prior to that, I'd be left to argue with receptionists who became *de facto* gatekeepers of my access to medical care that day. Some visits I'd have to sign or leave. Other days, a cheerful front desk person would look over my amended form and pronounce, "Don't worry about it—European patients always cross that one out, too."

Many subRosa art projects addressed the privatization of biology and the resistance against such enclosure—from the illegal acquisition of Henrietta Lacks' cancer cells by her medical provider (Johns Hopkins

Medicine 2025), to Jonas Salk's stated wish that the polio vaccine belong to everyone (Tan and Ponstein 2019).

Consent should be *sought* and not *presumed* when it comes to biometric and other data collection, storage, and sharing. The terms of such consent should be negotiated, impermanent, and apply to data as well as tissues.

Who is the steward to safeguard permanent information about individuals and populations? Is it for sale to the highest bidder? Leverage for international business deals? Are Clearview AI, Blackstone, and Amazon "too big to fail?" and what does that even mean?

Is the very concept of Stewardship—with its Abrahamic theological underpinnings—something to rethink?

Present data collection practices in the US seem to mimic the Intellectual Property laws surrounding biological materials: you can "discover" and own the DNA or the data of another being, but not your own. This is an artifact of Imperialism. Yet our personal data has great value and comes with great political and economic risk. It should belong to us, in perpetuity and by default.

Tech companies have accrued data and built communications infrastructure at a pace and volume that has far outstripped any nation-state's ability or will to regulate it, although the EU has been trying to keep pace. We should note with interest Elon Musk's meddling in global politics by turning on and off connections to his Starlink satellite network—a network on which many nations are reliant both to stage war and to resist it (Kim 2023).

As well, the privately held "NSO Group" has sold spyware to countries that have used it for counter terrorism, but also to hunt journalists, literally.

Palestinians and their diaspora have for decades been subject to nonconsensual surveillance experiments. They are the subjects of Israeli technology, but also Chinese technology developed to recognize Uyghur faces, then modified and sold for use in the occupied West Bank. In the ongoing retaliatory strikes following Hamas' October 7, 2023 attack on Israel, artificial intelligence has been deployed in an experimental fashion, including a program named *Lavender* that "assigns ratings to people in Gaza related to their suspected affiliation with Palestinian armed groups for purposes of labeling them as military targets," and another named *Where's Daddy* that "purports to determine when a target is in a particular location so they can be attacked there" (Human Rights Watch 2024).

What do the datapoints in your face reveal about your ethnic makeup or gender? As categorized by whom? And what of the Mestiza face? Was your family once German? Are you Hatian? Cuban? Syrian? Jewish? of unknown-to-yourself origins? What does your facial data reveal about your innermost thoughts and intentions? Should we indoctrinate a new kind of "poker face" to keep younger generations safe from state violence? Many women artists have already addressed the pressure from strangers to smile in public spaces. Facial recognition software has evolved to the point where it can now identify partially decayed faces, while still being rife with errors. UNESCO recently drafted a whitepaper *Recommending an Ethics of Neurotechnology* (UNESCO 2025). In it, a very smart international group is actively thinking through what to do about the new kinds of security and privacy issues that will arise with brain–computer interfaces. This includes the possibility of recording thoughts and emotions that one does not act upon, raising frightening new prospects for preemptive policing.

Media scholar Mirca Madianou closely examines how the United Nations is using new combinations of biometrics, artificial intelligence, and blockchain technology in the name of serving refugees efficiently—while simultaneously creating massive global databases of highly sensitive information. Madianou calls this the *Biometric Assemblage* and gives concrete examples being practiced today, where technical glitches in the *assemblage* can permanently harm cashless, stateless individuals confined in camps who rely on error-free digital transactions to obtain life-sustaining rations (Madianou 2019). To understand how this works, imagine buying a sack of flour with cryptocurrency, using facial recognition to verify the transaction, but the machine gets your identity wrong. It is not easy (and in some cases not possible) to sort out these errors, some of which may result in denial of access to basic services, or worse. Her scholarship further lays bare the need for a more complex understanding of the so-called global village.

Derma Twin

[Pittsburgh, 2024] ... As usual, my dermatologist has me sign a paper consent form AFTER she is done freezing off miscellaneous skin bumps with a hiss of dry ice during my annual exam. She zaps while her assistant listens for each "psssst" sound and keyboard-taps another hatch mark for the insurance claim. This is likely done in the name of efficiency and its twin benefits of serving more patients while generating more revenue: a literal *and* figurative gouging. It's the IRL version of a *shrink-wrap agreement* that only exists to protect our corporate overlords. *Psst, TAP, Psst, TAP...* On this visit, I ask to photograph the waiver—though I should not need to ask—and I explain my durational art practice, hoping not to come across as a litigious threat or mental health liability. She responds with curiosity and tells me that her employer is currently streamlining and digitizing the consent forms across the organization.

I mention Madianou's "Biometric Assemblage" and it resonates! She shares a story about her identical twin and being locked out of Facebook. One or both twins' biometric login caused the platform to presume an identity theft was underway and auto-blocked the account. The sisters found themselves struggling to prove their individual identities to Facebook and its parent corporation Meta. I wish I had asked for more details of how that was accomplished.

This problem is not limited to one platform, as I discovered later when trying to uncover Meta's requirements for proving oneself as a twin. As twinning rates rise globally, thanks in part to ARTs, frustrated parents have been complaining:

> *I tagged identical twin son 1 in a photo, and now all the pictures of twin brother 2 have twin brother 1's name. I can't seem to be able to undo this. Is there a solution?"*
>
> *[...] I have twin boys and constantly have to "*not this person*" in multiple pictures. [...] It's very frustrating for those of us who take pictures of identical persons and cannot get that photo to acknowledge two different people. I get it, my boys do look like each other. So then give me the option to "*choose*" the* People Album. (Apple User Forums 2025)

A few things are notable about these user complaints: here, parents are creating and commodifying a "data footprint" for their children without their consent; and the bad interface design is reshaping our language—to "not-this-person" becomes a verb. These frustrating "first-world problems" are another example of the aforementioned "User Pain Points." Madianou, however, digs more deeply into a very different type of pain that is inflicted when identity errors might generate a false criminal record.

While generative AI chatbots have already been released into the wild, we are also entering what some call the Agentic AI Era. This goes beyond chatbots that "generate" words and images from text-based prompts; it is AI that directs the actions of other AI, and it is being primed to make decisions with minimal or no human intervention.

Salesforce, a massive American e-commerce and AI company, breaks it down: "The future isn't about replacing human talent. It's about humans and agents working together in harmony" (Salesforce, 2025).

I have heard this idea of "collaboration" elsewhere—at Penn State New Kensington in workshops I have attended that prepare teachers to prepare their students for "jobs of the future," many of which are forecast to be about babysitting robots and preventing them from making very bad decisions (Penn State New Kensington 2023/2024). And Agentic AI—where robots babysit other robots—is already profoundly changing the landscape of modern warfare.

From *Shrink-Wrap Agreements* to *Consent Theater*

Why bother reading something no one was bothered enough to write is still excellent AI criticism. - Ursula Dorada (2023).

How might human cognition and social organization change if *ars memoriae* is yet again upended by machines, the latest which promise to do the work of idea generation, information recall, and decision-making for us?

Technology has come far and fast since the development of typography and it has a double edge: the Internet of Things for example is very vulnerable to security threats, while it extracts ever larger volumes of even more intimate biometric data. Psychological profiling and algorithms have become very powerful, anticipating our needs and wants, to keep us ALWAYS-ON-PLATFORM. Most who have this access enjoy great convenience while also struggling with very real device addiction.

Since 2016, I pad my time at airports so that I may pursue the time-consuming, literacy- and citizenship-dependent right to opt out of biometric data collection at US border crossings.

The US Transportation Security Administration (TSA) will tell you that biometrics was introduced in 2020, but the face scanners turned up at the Pittsburgh airport much earlier. Since the official rollout, you can find fine print on signage explaining that the facial scan is optional (Fig. 2.4). More prominent are the ads promising safety, speed, and a pledge that the photos are stored for a limited duration and purpose. This is an empty promise that ignores the fragile reality of data stewardship and trades on our presumptions about how photography and technology actually work.

I recently declined to have my face scanned at the Pittsburgh airport. The friendly TSA agent said "fine" and just scanned the face on my driver's license instead. On an overseas Delta flight a few years ago, I was encouraged to scan my FACE at the gate in lieu of my boarding pass, but I declined. I have a new car that, with some setup, I could start with my face, through my phone, but should I?

Thus, I've committed—when humanly possible—to the durational practice of taking the time to read and understand the Terms and Conditions that gatekeep most of my transactions with the Internet of Things and with apps. I want to normalize the discussion about what I find there. These things pop up at inopportune times, like mosquitoes. Like you, sometimes I do have to just click through and hope for the best, swatting and swiping my way through digital life.

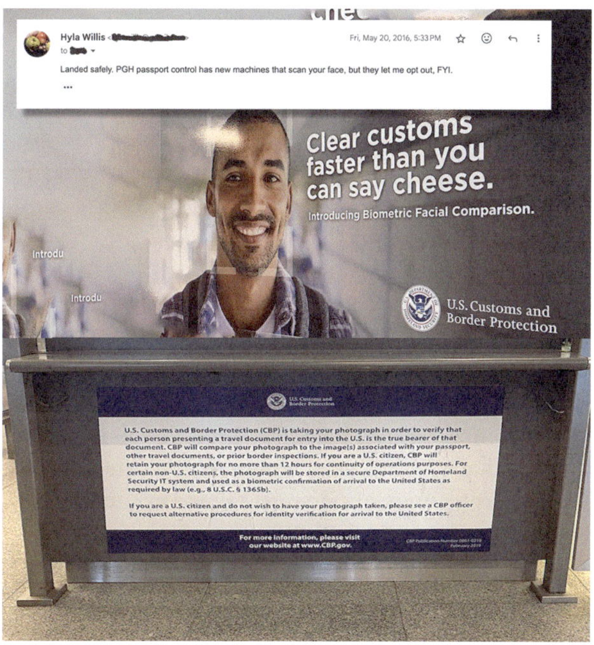

Fig. 2.4 A montage documenting the author's experience with the changing face of biometric security at US airports, 2016–2024.

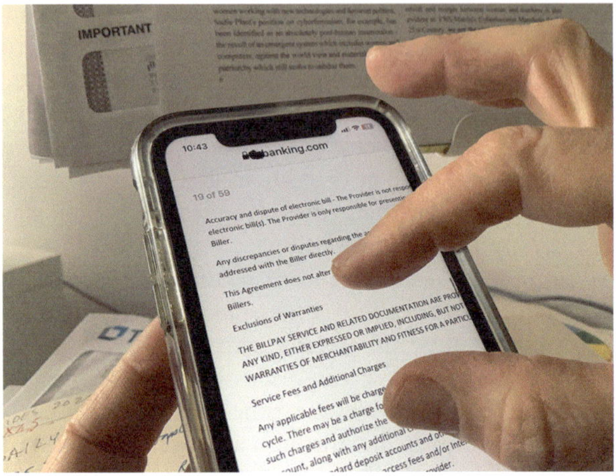

Fig. 2.5 A close-up of the 59-page update to Terms and Conditions required for Hyla Willis to use her online banking app.

I recently needed to check my bank balance and was confronted with a 59-page PDF policy update rendered in fixed-width typography too cumbersome to read on my smartphone screen (Fig. 2.5).

The term *shrink-wrap agreement* comes from the world of consumer packaging ...when you must tear open the plastic seal to find the agreement inside the box that admonishes that if you've broken the seal, you've agreed to the terms. At its heart, the shrink-wrap agreement is duplicitous. An inversion of this power dynamic might be considered shoplifting—a crime.

The cybersecurity and software development industries worry about a phenomenon called *consent fatigue.* They write papers about this and discuss it at conferences (Yacoob 2025). But let's call it what it is: *consent theater.* It is vital that we find ways to ameliorate the anxiety this causes and not succumb to feelings of personal failure in the face of a collective crisis. Choosing to slow down and read this stuff is awkward and inconvenient. I worry about making myself a target or just coming across as an asshole—the quiet vegan cousin at Thanksgiving Dinner whose personal choice conflagrates family rage, or worse: a *contrarian woman.*

However, talking with people about terms and conditions, privacy, and identity, and chatting up the enforcers about their working conditions...

has led to very interesting moments of commiseration. These are important steps toward imagining the future that we would rather have together.

There's an invisible fence surrounding our bodies and our attention that is too large and fast-changing to contain. It is hard to even understand if it has a centrality or from where it emanates. "The Cloud" is a growing public/private mesh of satellites and climate-destroying data centers that cause harm through air, water, and sound pollution.

Collectively, we need to feel for the holes in this invisible fence, to stretch the gaps like hungry garden bunnies. By naming and envisioning the containment, we might seek or create the refugia that we can inhabit together.

If you're feeling curious but overwhelmed by this prospect, here's a place to begin:

May I recommend that you find a friend and spend time reading the Terms and Conditions and Privacy Policy for the UBER app? Do it together and discuss what you find.

You'll come to understand what kind of company UBER is not—it is fundamentally NOT a transportation provider. And you'll discover what rights you give up, in perpetuity, into the future as well as the past, forever and ever, even when the terms change, which may happen at any time.

Next, type the words PIZZA + UBER + LAWSUIT into a search engine. There are material consequences to your choice to click OKAY that may surprise or shock you. Then, take a page from subRosa and others who, through art, philosophy, law, and policy, are critical of the concept that "your choice = your responsibility" (Cooper Law 2025).

The Slow Art of Reading the Terms: A Call to Reverse Engineer the *Biometric Assemblage*

It's true that we're afforded many conveniences for a price that most of us will not fully understand, and not because we are stupid or lazy. Opting out, or *the slow art of reading the terms*, may feel futile, but it is also a form of literacy with which we should engage. And you don't need to do it alone.

I've learned from my university colleagues that human psychology and not shoddy coding is the greatest vulnerability when it comes to cybersecurity and protecting data privacy. This makes me oddly optimistic that a mindset shift is possible, that a "slow food movement" approach to "fast food tech" could gain traction and bring solidarity, joy, and a greater quality of life. And art can always bring possibilities and conviviality to this effort. One possibility is simply holding a place in the public imagination that things can operate differently than they currently do. There is an opportunity for solidarity and merrymaking if we share close readings of these texts and how they impact our lives, a rallying of energy to reject and deflect sludge and dark patterns. We might also feel empowered to stop demanding heedless downloads from each other.

In the early 1990s, my coworker Gillian and I were robbed at gunpoint while walking home from a concert in Seattle. We returned to the venue to use their phone. I filed a police report and called the bank during their business hours to cancel the series of checks the thieves took. That should have been enough, but sloppy retailers (or perhaps cashiers in cahoots with the robber) allowed someone to use the checks to purchase jewelry and electronics in my name. For two long years, I was chased by debt collectors and soon learned that "bad debt" (such as when a store accepts checks that they know are phony) gets bundled and auctioned to the next-sleaziest bottom feeder who begins the debt collection process all over again. I was finally released from this *doom loop* by a human. After trying to buy tires and getting rejected once again by the TeleCheck service, I asked the cashier to lend me their telephone to call the reporting agency. Once connected, I explained, shouted, reasoned, and finally—*I cried.* My sobs cut something loose in the human being who took my call, and she wiped away my troubles immediately with just a few audible keystrokes. Perhaps this is why there is movement aground for the *right to talk to a human* in customer service transactions.

In an Op-Ed segment, CBS News anchor John Dickerson declares that "bad customer service can lead to moral collapse." He describes being caught in a frustrating doom loop with his bank's online chatbot while trying to clear up an erroneous fee to his account. At a certain point, he realizes that the AI has no interest in resolving the issue, and he decides to lie about the true purpose of the call, dangling information that he knows the bank will want from him. It works, and the bot offloads the chat to a real human, but Dickerson clearly feels compromised (Dickerson 2025).

In an essay on sludge in *The Atlantic*, Chris Colin describes "Admin Night"—a social hour at his house where friends hang out together with their laptops and tick through personal administrative tasks online, or socialize while on hold for lengthy customer service calls. "Learning of sludge's existence, [Stephanie] Thum, the bureaucracy researcher, told me, is the first step in fighting it, and in pushing back against the despair it provokes." (Colin 2025). While Admin Night has a tinge of privileged cuteness about it, I am strongly in favor of nearly any form of solidarity that helps build understanding of what is taking place within these dark patterns and generates creative ideas about pushing back. For many individuals, the stakes are quite higher than inconvenience.

Y.K. Hong, a community organizer, artist, speaker, and coach who offers specialized training in anti-oppression and "tech justice," maintains an active social media presence where they post instructional videos of themselves opting out of facial recognition at various airport checkpoints. Through their online platform, they offer a subscription for detailed instructions on how to opt out and/or minimize data harvesting by Big Tech. Hong is also a fierce online advocate for "demanding a culture of consent" when it comes to the scraping of data and of creativity to train AI (Hong 2025).

Are personal attempts to "opt out" of the myriad Terms and Conditions and other encroachments on our data bodies mere exercises in futility and privilege flexing, or might they foster digital literacy leading to new forms of creative resistance to the sludge? How might we co-author our own End User License Agreements within the atmosphere of pervasive information-gathering and -sharing as a way of grappling with its impact on our privacy, livelihood, personhood, and sense of oneself?

Allegations of Failure and a Call for Creative Experimentation

Journalist Masha Gessen, ruminating in a 2017 lecture during Trump's first term, proposes that his ascendency also represents a failure of the imagination on the part of the opposition. Gessen argues that not doing the work of imagining a Trump victory put the left on its back foot, positioning that it only temporarily recovered from (Gessen 2017). Within the first few weeks of Trump 2.0, the work has begun to normalize (in our imaginations, since the law still forbids it) the possibility of Trump 3.0.

For those unhappy about Trump's ascendancy, it is perhaps too easy to point to one another with allegations of failed imagination (does no one on the left watch *The Simpsons*?). Equally troublesome, though, is to view change through a universally dystopic lens. We must do the work of both—while a project like *SmartMom* does the necessary work of imagining dark patterns and can give a reprieve through humor, we must also imagine new forms of mutual aid and beneficial technology—the Terms and Conditions we would rather live by. Several of subRosa's projects make attempts at this through playful, collaborative engagement with the audience (see Chaps. 4 and 6).

Kill Your Apologies

Consent has great value, comes with great risk, contains nuances, and should be ours to give.

Consider divesting in notions of giving and receiving apologies and thanks with the coming AI chatbots. The performative civility returned by machines just usurps more human psychology at great environmental and psychological cost. When faced with seemingly endless extraction & enclosure, joyfully seek and demand repair and restitution instead of politeness and urgency from technology companies. The Terms and Conditions seek only to leverage your anxiety and peel away your rights and resources.

If a central question of this Bio-Art Series is "What happens when we treat life as a raw material for artistic expressions?" Let this essay further ask, what happens when biological and biometric material is conceived of as a system of discrete units not unlike an alphabet that can be mechanized like typography?

Performing a close reading of Terms and Conditions is an unexpected response and a form of experimental creative rebellion when communicated with others. The fact that you bother to notice and read them will be a shock. Sharing details about their length and content is a further shock. It is also a refusal to be confined by typography, to let the alphabet be turned against us.

It is time to return the sludge and OPT OUT.

References

Adobe Fonts. 2018. "Adobe Jenson," retrieved 08-03-25. https://fonts.adobe.com/fonts/adobe-jenson#about-section

Apple User Forums. 2025. "Photo face recognition - identical twins thread". https://discussions.apple.com/thread/252184460

Bringhurst, Robert. 1996. *The Elements of Typographic Style*. Hartley & Marks Publishers.

Colin, Chris. 2025. "That dropped call with customer service? It was on Purpose." *The Atlantic,* June 29. https://www.theatlantic.com/ideas/archive/2025/06/customer-service-sludge/683340/
Cooper Law. 2025. "Cooper Law Whitepaper: The Impact of Uber's Arbitration Clause on Passenger Rights in Personal Injury Cases." https://www.cooper-law.com/blog/the-impact-of-ubers-arbitration-clause-on-passenger-rights-in-personal-injury-cases/
Dave, Paresh and Dastin, Jeffrey. "Exclusive: Ukraine has started using Clearview AI's facial recognition during war" March 14, 2022 Reuters. https://www.reuters.com/technology/exclusive-ukraine-has-started-using-clearview-ais-facial-recognition-during-war-2022-03-13/
Dickenson, John. 2025. "Bad Customer Service Can Lead to Moral Collapse." *Reporter's Notebooks*, July 28. CBS News.
DiStefano, MJ, JM Markell, CC Doherty, GC Alexander, andGF Anderson. 2023. "Association Between Drug Characteristics and Manufacturer Spending on Direct-to-Consumer Advertising." *JAMA,* 329 (5): 386–392. doi:10.1001/jama.2022.23968.
Dorada, Ursula. 2023. Twitter Post. (Now X), Nov 15. https://x.com/sulamoon/status/1724888928866476417
Federal Trade Commission. 2022. "FTC Report Shows Rise in Sophisticated Dark Patterns Designed to Trick and Trap Consumers." Press Release, Sept 15. https://www.ftc.gov/news-events/news/press-releases/2022/09/ftc-report-shows-rise-sophisticated-dark-patterns-designed-trick-trap-consumers
Fierce Biotech. 2007. "Viragen dumps transgenics program, cancels pacts." June 22. https://www.fiercebiotech.com/biotech/viragen-dumps-transgenics-program-cancels-pacts
Gessen, Masha. 2017. "How We Survive and Autocracy." A recorded e-flux lecture, May 24. https://www.e-flux.com/podcasts/406720/masha-gessen-on-how-to-survive-an-autocracy
Gibbons, Sarah. 2021. "Three Levels of Pain Points in Customer Experience." Neilson Norman Group, May 16. https://www.nngroup.com/articles/pain-points/
Hill, Kashmir, and Corey Kilgannon. 2022. "Madison Square Garden Uses Facial Recognition to Ban Its Owner's Enemies." *The New York Times*, December 22 (updated Jan. 3, 2023). https://www.nytimes.com/2022/12/22/nyregion/madison-square-garden-facial-recognition.html?unlocked_article_code=1.9U8.9tLs.32FH4g4Csn-c&smid=url-share
Hill, Kashmir. 2020. "The Secretive Company That Might End Privacy as We Know It." *The New York Times*, January 18, 2020 (updated Nov 2, 2021). https://www.nytimes.com/2020/01/18/technology/clearview-privacy-facial-recognition.html?unlocked_article_code=1.9U8.lNdJ.zJRX9XwArhuQ&smid=url-share

Hong, Y.K.. 2025. "Know Your Opt Out and Privacy Clauses." https://keepbeyond.com/optout/

Human Rights Watch. 2024. "Questions and Answers: Israeli Military's Use of Digital Tools in Gaza," September 10. https://www.hrw.org/news/2024/09/10/questions-and-answers-israeli-militarys-use-digital-tools-gaza

Jaret, Peter. 2018. "Exposing vulnerabilities: How hackers could target your medical devices." *AAMC News*, November 12. https://www.aamc.org/news/exposing-vulnerabilities-how-hackers-could-target-your-medical-devices

Jardiance Commercia. 2024. *Commercial Archivist You Tube Channel.* https://www.youtube.com/watch?v=AMHMU0YPFEc

Johns Hopkins Medicine. 2025. "The Legacy of Henrietta Lacks." https://www.hopkinsmedicine.org/henrietta-lacks

Kim, Victoria. 2023. "Elon Musk Acknowledges Withholding Satellite Service to Thwart Ukrainian Attack." *The New York Times,* September 8. https://www.nytimes.com/2023/09/08/world/europe/elon-musk-starlink-ukraine.html

Kingsolver, Barbara. 2005. "Remarks at the Opening of Higher Ground Recovery Residence." https://www.theguardian.com/books/2025/jul/05/the-damage-is-terrifying-barbara-kingsolver-on-trump-rural-america-and-the-recovery-home-funded-by-her-hit-novel

Lessig, Lawrence. 2004. *Free Culture: How Big Media Uses Technology and the Law to Lock Down Culture and Control Creativity.* Penguin.

Madianou, M. 2019. "The Biometric Assemblage: Surveillance, Experimentation, Profit, and the Measuring of Refugee Bodies." *Television & New Media*, 20 (6).

McGowan, Kathleen. 1999. "The Welfare Estate." *City Limits,* June 1. https://citylimits.org/the-welfare-estate/

McLuhan, Marshall. 2011 (1962). *The Gutenberg Galaxy: The Making of Typographic Man.* University of Toronto Press.

McLuhan, Marshall. 1964. *Understanding Media: The Extensions of Man.* McGraw-Hill.

Monotype. 2019. "Fonts: The underrated political campaign staffer." https://www.monotype.com/resources/expertise/underrated-political-campaign-staffer

Mountainhead, 2025. Directed by Jesse Armstrong, HBO.

NBCU Academy. 2024. "Chyrons, Lower Thirds, Banners: What Are They?" February 21. https://nbcuacademy.com/chyrons-lower-thirds/

Penn State New Kensington. 2023 and 2024. "Future of Work Faculty Academy." Workshops. https://newkensington.psu.edu/future-work-events-faculty

Pennsylvania Attorney General. 2025. "Your Right to Rescind." https://www.attorneygeneral.gov/protect-yourself/consumer-advisories/your-right-to-rescind/

Salesforce YouTube Channel. 2025. "The Agentic AI Era | Salesforce AI Research." https://www.youtube.com/watch?v=D6zHWQP15K8

Schwartz, Hunter. Dec 2, 2022. "Trump's 2024 brand sticks with what voters know. Can it work?" At https://www.yellopolitics.com/p/trumps-2024-brand-sticks-with-what

Sharma, Sarah, and Rianka Singh, eds. 2022. *Re-Understanding Media: Feminist Extensions of Marshall McLuhan*. Duke University Press.

Tan SY, Ponstein N. "Jonas Salk (1914-1995): A vaccine against polio." *Singapore Medical Journal*. 2019 Jan;60(1):9–10. doi: 10.11622/smedj.2019002.

UNESCO Digital Library. 2025. "Draft Recommendation on the Ethics of Neurotechnology." https://unesdoc.unesco.org/ark:/48223/pf0000394866

Wegovy Commercial #2. 2024 and 2025. *Commercial Archivist You Tube Channel*. https://youtu.be/QnUUVesrRxQ?si=hRRuHeBX7jSrDF0c

Willis, Hyla. 2025. Personal Digital Archives.

Yacoob, Safna Y. 2025. "Consent Fatigue: What is it and What Should Businesses Do?". *CookieYes*. Blog Post, June 11. https://www.cookieyes.com/blog/consent-fatigue/

CHAPTER 3

Domain Errors! A Virtual Roundtable

Hyla Willis and Irina Aristarkhova, with contributions from María Fernández, Amelia G Jones, Tania Kupczak, Laleh Mehran, Susanna Paasonen, Nell Tenhaaf, and Michelle M. Wright

Preface

In the fall of 1999, subRosa began work on an experimental anthology that became *Domain Errors! Cyberfeminist Practice*s (Autonomedia 2002 / 2003). In the book's Introduction, editors María Fernández, Faith Wilding, and Michelle M. Wright stated their intention was "to initiate a feminist and postcolonial critique of embodiment, difference, and racial prejudice within cyberspace, biotechnologies and cyberfeminism... We wanted the contents of the project to go beyond general critiques... to include areas repressed in previous cyberfeminist discourse and criticism... [and are] committed to seeking out contributions by women whose voices had not been heard in the cyberfeminist context" (9).

Now, over twenty years later, we—Hyla Willis and Irina Aristarkhova—wanted to revisit *Domain Errors!* as one of subRosa's legacies. We sought reflective responses from its original authors and subRosa members to the questions the book asked, the tactics it proposed, and how the contributors' relationship to cyberfeminism(s) has changed. In their responses, María Fernández, Michelle M. Wright, Laleh Mehran, Susanna Paasonen,

H. Willis et al., *Night Sweats: Cyberfeminist Practices*,
Palgrave BioArt,
https://doi.org/10.1007/978-3-032-06955-9_3

Amelia Jones, Nell Tenhaaf, and Tania Kupczak offered their insights and updates to their creative, activist, pedagogical, and scholarly interests. We assembled responses according to the questions that we sent out and created a Virtual Roundtable.

As author-contributors to *Domain Errors!* (and Willis also was the book's designer and typesetter), we were excited to reread the book at this moment in time, when not only the worlds of Cyberfeminism, BioArt, and Tactical Media have changed, but also the work of its many authors and creators who have gone on to make significant contributions to intersectional feminist and media theory, art history, pedagogy, art, and activism. In a May 2024 email, we wrote: "We [Aristarkhova and Willis] are now spearheading a publication that examines the legacy of the subRosa art collective. We are writing to you today because of your involvement in one of subRosa's important projects: the book *Domain Errors! Cyberfeminist Practices.* ... As we re-read the book over the two decades after its publication, we hope you will join us in revisiting some of the questions it asked and the tactics that it proposed."

What you will read below are Seven Questions that were posed to the book's contributors and subRosa members, followed by the complete text of the replies from all who were able to respond. Knowing that most—but not all—of the book's contributors spent the intervening years teaching in postsecondary institutions within the United States, we also sought to capture their thoughts on interactions with younger scholars in a changing landscape for higher education. These have become our key terms in this chapter, reflecting the topics covered:

Intersectionality, reproductive justice, reproductive labor, art strategies, solidarity tactics, feminist pedagogy, feminist scholarship, cybernetics, pornography, menopause, backlash, LGBTQIA+, DEI.

In our questions, we sought to capture both subRosa's and *Domain Errors!* legacies and future-oriented topics pertinent today. Our questions, in a summary form, were:

1. **How might we persist with intersectionality during the current backlash?**
2. **What do you find the most helpful in this current moment of research and creative work around gender, sexuality, and parenting?**
3. **What do you think of this current moment in relation to reproductive labor, justice, and technoscience?**
4. **What art and solidarity tactics inspire you the most right now in your scholarship, creative work, and/or teaching?**

5. **How has your teaching of cyberfeminism, technoscience, or sexuality and gender-related topics evolved in relation to changes in your students, your local context, or your own thinking/ making? What are you teaching right now?**
6. **Are there key texts or works of art that have significantly reshaped your work and thinking in the 20+ years that followed Domain Errors? What are you currently reading or researching, and what is the most interesting to you?**
7. **Is there anything else you would like to respond to, or wish we had asked you about?**

In their wide-ranging responses to these questions, our contributors spoke of new art and scholarly works, of new activist practices and collectives that sustain them and their work. They have also shared many challenging experiences and complex questions around Diversity, Equity, and Inclusion programs, their institutionalization, and the current backlash against them. They shared why they think we live in an anti-progressive moment in general, and where reproductive technologies and global gender panics factor in. Our Roundtable contributors provided inspirational, inspired, and inspiring, references to works of scholarly and creative output that we hope will be treasured by our readers, to many examples that nourish their own writing, thinking, and making, and they were generous to share what they are working on right now. It is fascinating for us to recognize in this Roundtable an affinity among members of this community, especially in their commitment to be the agents of transformative change in academia, art, and activism. We are very grateful to our contributors for their generosity, which you will encounter in this Virtual Roundtable. Since the questions and responses are so wide-ranging, the reader can choose to focus on separate questions or read them sequentially. In the spirit of *Domain Errors!*, we wanted to retain this range of individual formal choices and rhetorical styles among our respondents, without imposing our own uniform narrative.

In this Virtual Roundtable endeavor, we also wanted to pay tribute to the work that *Domain Errors!* represented as being more than a traditional publication, reflecting subRosa's ethos, its values, and core principles in a book form. In 2002, with grant support from the Creative Capital Foundation and the STUDIO for Creative Inquiry at Carnegie Mellon, subRosa was able to share production costs with Autonomedia and retain open access distribution, which was very important to the collective, even in those early times.

The deliberate choice to work with an independent, non-academic, non-profit publisher was perhaps a risky calculation for some of the authors who were in the early stages of tenure-track career paths: however, it allowed the group to treat the book as an extension of its art practice, and for the editors to include a wide variety of approaches and thought, and funding for the book was competitively obtained. This was a curated collection of contributions, bypassing the traditional open call and editorial processes of an academic press. *Domain Errors!*— as the front cover states—is "a subRosa project."

The contributions were international and intersectional in scope and included experimental layouts and writing alongside deeply personal narratives about racism and fertility technologies; humorous pages, rants, manifestoes, and historic timelines on serious topics. It is fascinating now to realize that Nell Tenhaaf's work (Fernández et al. *Domain Errors!* 2002, 251) exploring viewer empathy through a custom algorithm was happening before most of us were holding smartphones or using social media. "The Refugia Manifesto for Becoming Autonomous Zones [BAZ]" (Fernández et al. 2002, 277) was written before the Second Gulf and Syrian Civil wars displaced millions of people—its title a nod to a specific term used in climatology and agriculture and a play on the title of Autonomedia's best-selling *Temporary Autonomous Zones* [T.A.Z.] book by Hakim Bey. Autonomedia also published works by Silvia Federici that were important to subRosa.

The book cover, too, contains Easter eggs that might now be read as hauntings (see Fig. 3.1): a satellite image of the Chesapeake Bay with an inset of a home that has been crushed by a natural disaster, both images sourced from the National Oceanic and Atmospheric Administration, a US government organization currently facing devastating cuts and the disappearance of its datasets.

Individual chapters were made available on the subRosa website, making *Domain Errors!* freely accessible to individuals and educators, including a translation to Spanish by Gabriela Adelstein of Fernández and Wilding's "Situating Cyberfeminisms" chapter.

subRosa hatched the idea for *Domain Errors!* following a summer reading intensive in 1999 (Fernández et al. 2002, 9). The editors outlined their structure in the book's introductory chapter:

> *Domain Errors!* is divided into three sections that combine theory, practices, critique, and artist projects. Section One, "Racism and Cyberfeminism in

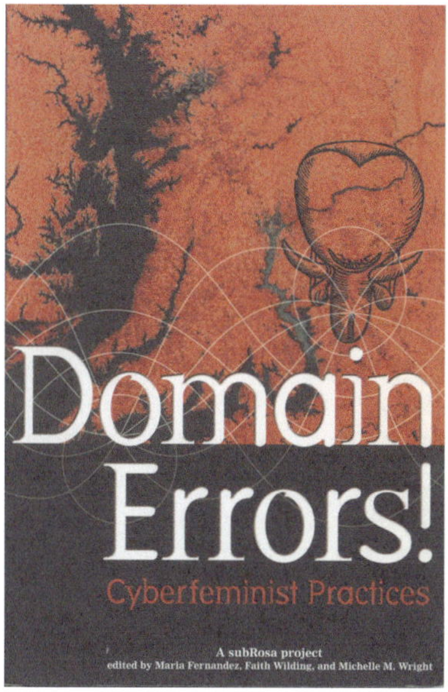

Fig. 3.1 The front cover design of *Domain Errors! Cyberfeminist Practices* (subRosa and Autonomedia, 2003).

> the Integrated Circuit," discusses lived aspects of the intersections of race, gender, and technology in the situations of women in different countries and cultures of technology. … Section Two, "The Female Flesh Commodities Lab," offers searching discussions of the molecular invasion and colonization of the female body by Assisted Reproductive Technologies (ARTs), recombinant genetics, and new medical and surgical flesh technologies such as aesthetic vulvar surgery. … Section Three, "Research! Action! Embodiment! Conviviality!" takes a hands-on approach to theoretical and embodied cyberfeminist practices. It offers a mix of activist and artist pages and projects. (Fernández et al. 2002, 11–12)

The *Domain Errors!* editors also left room for the then-nascent subRosa art collective to summarize its earliest works—including the performances "Sex and Gender Ed in the Biotech Century," "Expo EmmaGenics," and "U.S. Grade AAA Premium Eggs"—ending in April 2002 (subRosa 2002, 217–221). subRosa's essay "Stolen Rhetoric: The Appropriation of Choice by the ART Industry" (subRosa 2002, 135)—originally published in German in Kunstforum International—is republished in translation in Section Two. Following the publication of *Domain Errors!*, it was later given a new illustrated layout and published in n.paradoxa and as a free-standing pamphlet handed out at performances and conferences.

The book strove to be a multifaceted resource. subRosa members compiled a "selected bibliography" (ibid., 285) organized by themes [New Reproductive Technologies; Sex/Gender/Body/Technology; Medicalization/medicalized Representation of Sex and Gender; Queer Theory, Feminist Theory, Postcolonial Theory; and Cyberfeminism/Science/Gender/Technology] and reflective of the range and topics that had been shared in its beginning days as a reading and discussion group.

Collective members, editors, contributors, and Autonomedia agreed to publish the book with an "Anti-copyright" notice at the front, to encourage and enable the sharing and spreading of the ideas and information that it contained.

All of this is listed here as a testimony to the original intentions and their formal materializations in *Domain Errors!* as one more example of the multitude of subRosa's cyberfeminist tactics and strategies. It is our hope that the Virtual Roundtable that follows below speaks to a legacy of that original ethos.

Q1: Centering Intersectionality During a Backlash

Today, even the least critical champions of algorithms, machine learning, and new media art acknowledge the problems of racist, ethnocentric, and gender biases in data science, data procurement, and AI.

In 2002, *Domain Errors! Cyberfeminist Practices* centered identity politics, inequality, racism, (settler) colonialism, and imperialism in its discussion of cyberfeminism. Editors María Fernández, Michelle M. Wright, and Faith Wilding opened the book with a section—"Racism and Cyberfeminism in the Integrated Circuit"—devoted to these topics that included their own contributions alongside chapters from Lisa Nakamura, Radhika Gajjala, Annapurna Mamidipudi, Susanna Paasonen, and Irina Aristarkhova.

Today, Legacy Russell, Mindy Seu, and Stephanie Dinkins—to name a few—challenge biases in new media art, design, and computer science, through writing and creative work that touches upon critical data studies, for example, and offers new paradigms of thinking and critique. Their work takes place during a time of intense backlash against Diversity,

Equity, and Inclusion initiatives in academia, the art world, and in techno-scientific industries.

How Might We Persist with Intersectionality During the Current Backlash?

Tania Kupczak
Intersectionality has become a buzzword in the same way that DEI can be used as virtue-signaling in white-led spaces. I believe the work of intersectionality requires a low barrier for entry—no gatekeeping or over-intellectualizing—and a high bar for accountability and behavior. Those of us with privilege are required to first acknowledge it and then set aside personal discomfort to leverage that privilege for folks who aren't safe doing so. It may be because I am steeped in mutual aid movements that I don't see a way forward without intersectionality as a core principle of whatever work we put our hands to. For me, it's a practice of always holding this value in my head and asking how a project, conversation, or piece of art uplifts this value.

María Fernández
Feminism and diversity enjoyed widespread acceptance for a brief period. In the last hundred years, both confronted resolute repression and waves of opposition. Think of the "backlash" of the 1990s (Susan Faludi 1991). Diversity was undervalued for most of the last century. Even within feminist and cyberfeminist movements, it took a long time for issues concerning race and difference to be taken seriously. The only way intersectional feminisms will persist is if we continue to do our work in whatever way we can. This will require the ingenuity and creativity that feminism always involves. Let's work with what we have.

Amelia Jones
Even as a late "boomer" (b. 1961), my relationship to intersectionality has been transformed since I first encountered the concept in Kimberlé Crenshaw's work in the 1990s. My earlier (1992) well-meaning but additive approach as a feminist to the complexity of identities was transformed when I moved to Manchester, UK, and then to Montréal, Canada, and experienced the oddness and disjunctiveness of my own linguistic, class, ethnic, national, and sex/gender identifications. My book *Seeing Differently: A History and Theory of Identification and the Visual Arts*

(2012), which explores the complex history of identity politics in relation to visual culture, marks that shift. In the book, I seek to inhabit my own intersectionality more profoundly and to find more effective ways to articulate this complexity across the board as I write about and curate visual and performance art work.

Younger generations inhabit intersectionality from the get-go, particularly obviously in terms of the increasingly visible non-normative gender/sex articulations among educated youth of all kinds, especially in the more Westernized parts of the world (i.e., Europe and its colonies or former colonies, especially those of Anglophone cultural formation). Until recently, these articulations were fairly well supported by neoliberal Western governments. The current backlash is violent and distressing, but totally familiar to anyone who has studied the history of Western culture over the past 100 years. We are now effectively in a time paralleling the 1930s, wherein (especially in the case of Germany, coming out of the permissive Weimar period) a huge backlash against the gender/sex freedoms and the previously liberal policies towards folks across the spectrum of identifications shifted Western countries into fascism.

Capitalist and late-capitalist structures of these societies produce inequities, in turn, social instabilities, and thus generalized anxiety. This generalized anxiety is the backdrop to backlashes against increased rights; opportunistic billionaires and politicians—usually older white men, still—seek to regain power by hate-mongering against those they cast as "threatening." This is a tried and true method of garnering support for these seemingly more "populist" voices (in the US, Musk, Trump, Bannon, etc.). Given our addiction to extractive (late-) capitalism, the instabilities and thus the reversion to intolerance and attempts to reestablish control over bodies and over societal ideas about identity—including perhaps especially those fragile concepts circulating around gender and sexuality—are perhaps inevitable. It certainly feels that way right now.

Laleh Mehran

Advancing intersectionality through DEIJ (Diversity, Equity, Inclusion, and Justice) in an increasingly hostile political climate first requires confronting the ways we have not always supported one another in meaningful, structural ways. While tokenism has evolved alongside cultural recognition, institutions largely remain unwilling or unable to fully acknowledge the systemic inequities they have created and continue to uphold.

Whitewashing and selective forms of support often obscure deeper, more deceptive actions that actively undermine true equity.

Rather than fostering solidarity, competition over limited resources exacerbates divisions among underrepresented groups, reinforcing cycles of minoritization, isolation, and hierarchical class dynamics within creative and academic spaces. My anticipation of joining communities of people of color and other marginalized populations has often collapsed when my nationality is revealed; when I am not part of the majority or of a group that is of strategic interest to the institution, I am met with exclusion rather than solidarity.

Though these challenges may seem evident, they bear repeating—along with an honest reckoning that we have not always done enough to practice full intersectionality. Recognizing the significant labor already undertaken is essential, as is acknowledging that these efforts will only grow more difficult. For intersectionality to be truly transformative, it must go beyond representation and be actively practiced through knowledge-sharing without gatekeeping, allowing diverse perspectives to shape discourse and influence decision-making. This also requires those with greater visibility and recognition to take an active role in advocacy, expanding opportunities, stepping aside to create space for others, ensuring collaborators receive equal recognition, and insisting on diverse representation in professional and creative spaces. It can include refusing participation in engagements where organizations prioritize performative gestures over genuine DEIJ commitments.

Educational and cultural institutions are increasingly reluctant or unable to support marginalized scholarship and the creative arts, often fearing funding loss or alienating benefactors. Whether driven by self-preservation, a lack of personal investment, or a means to uphold systemic racism and exclusion, this hesitation reinforces structural inequities. I've witnessed requests to "tone down" critical content in exhibitions as a condition for securing a platform, making it clear that those who control funding have the power to decide what is deemed acceptable for public engagement.

Educators and artists have long demonstrated intersectionality in practice, both within and beyond academia, through methodologies such as fostering interdisciplinary collaborations with technologists to bridge critical race theory, feminist technoscience, and digital art, driving systemic change. Public-facing interventions in my classes have been an expansive experience for the students, the audience, and me. By integrating technology with established fields like archaeology, these projects have critically engaged various publics with pressing issues such as immigration. Many students have gone on to apply these methods in their own work, using technology and interdisciplinary approaches to address social issues in their creative practices or careers dedicated to public good.

There are thinkers and makers who actively disrupt inequities, demanding accountability and forging new paths for change. However, such efforts require protection from various forms of retaliation, ranging from tenure denial to forced deportation, and must be supported by broader and more diverse demographics. Access remains a critical issue; not all creatives, scholars, and activists who seek to participate in these conversations have the necessary resources or platforms to amplify their voices and push back against growing hostility. In response, collective strategies must extend beyond individual visibility. This includes building platforms without always centering oneself as the figurehead, fostering open-source knowledge, strengthening digital solidarity, and leveraging decentralized networks for resistance and the redistribution of power.

Despite efforts by trolls and bots to disrupt and destroy online spaces for marginalized populations, strong communities continue to form, mobilizing through decentralized networks, encrypted communication, and alternative platforms such as Mastodon, Bluesky, PeerTube, and Solidarity Economy networks. In this climate, ensuring that people do not feel alone, helpless, or paralyzed is more critical than ever.

Susanna Paasonen

There is no other way, really, than to keep at it, even as it's important to do individual and collective care work so that things remain viable. And there are critical masses of people in and across the arts and academia doing this work despite the circumstances, which helps, as these people are also networked.

It's my sense that intersectionality wasn't particularly popular within the cyberfeminist discourse of the 1990s, to which the book then made both a contribution to and an intervention in. To generalize a little, the playfulness of cyberfeminist tactics also involved a certain evasiveness in terms of how feminism, and the different strands and legacies of feminist thought, were understood, to the point that it wasn't always clear what the feminism in cyberfeminism pointed to, or indeed how it was feminist/did feminism. This also came with a certain disinterest in, or blindness towards, embodied differences and social relations of power beyond a kind of gender binary ("the future is female" is a great slogan, but also isn't).

It's also important to note that the current backlash plays out in different ways internationally. For example, in my own work context (Finland), it's only the past few years that structural racism has become, if not always acknowledged, then at least much publicly discussed in ways adding urgency to intersectional approaches, so that the backlash overlaps with an overall realization that things are, indeed, bad. Very slow going, up here, and the rise of the populist right has definitely been drastic. But I guess my take on the question is that things weren't particularly great 20+ years ago and while they're certainly not great now, either, we're in a different place, also in a good sense, when considering what questions and approaches have become standard in critical humanities and social sciences inquiry, and the volume of work that is out there for us to draw on.

Nell Tenhaaf

Generative AI right now is too much like this: the past is chewed up in the service of a utopian and rather mindless idea of the future. It's a huge impersonal monster that has a tendency to mash up all past human experience into pap. One might ask, what else is new? Norbert Wiener (1948) warned in the 1940s about a flaw in the learning machine if it is looking for 'social efficacy'. And now we can see how learning machines built on the past can reinstate social problems rather than find the way forward out of them. It's a rhetoric that's too much like that of political demagogues who misrepresent the past so as to make false promises, when what we need is a considered concept of the future that studies the past, including humanity's many repeated historical mistakes. If Large Language Models (LLMs) are only set up to take what we know and what we've done and spit out a composite of the 'nothing new' from all of it, this is at the expense of the larger question that aware humans have always faced – whose future are we interested in? There has to be a sense of a path forward that we are hoping to forge for the whole thing to make any sense. I think that in the arts, we're seeing the beginnings of making AIs more personalized and individuated by setting training models to work vigorously at approaching past experience differently, to reveal it rather than ignore it. Stephanie Dinkins' work is a good example, with its focus on a considered revisiting of memory/history to tease out pathways to understanding.

Michelle M. Wright

I think there are a number of ways to persist with intersectional analyses, and we should probably try almost all of them, but when it comes to academic discourses, I personally would plump for weaving class and other forms of socioeconomics into our discussions. The two models for intersectional analysis that we tend to deploy is either, first, a white supremacist norm in Women's, Gender, and Sexuality Studies, meaning most scholars who work on white feminism want to keep racial analyses sequestered, and only brought out when it suits them, allowing their whiteness to remain a value of implicit, positive normativity rather than an active privilege always seeking to operate as invisibly and as centrally as possible. Second, in Black Studies (I am woefully undereducated in indigenous studies, Middle Eastern and North African Studies, Asian American Studies, Arab American Studies, Latinx/Latine, Chicano/Chicana Studies), our model is Black heteropatriarchy with a bourgeois normative—that is, the vast majority of the Black chattering class, both within and outside academia, are economically elite and view any sort of discussion of economics as a dangerous distraction.

Sadly, we are equally defensive around issues of gender equality and sexual tolerance, and tend only to promote those scholars who pretend that Black nationalism and/or Pan-Africanism are as inclusive as they need to be when it comes to Black women and queers.

I think we all have work to do, and I include myself. We need to assume that no matter how hard we try, we are never wholly intersectional, and we need to acknowledge those shortcomings in our writing. I believe that, at its root, all academics tend to orient themselves within Enlightenment norms, where only the wealthy and otherwise socioeconomically privileged had access to the time, power, and audiences to discuss matters such as equality and justice. That comes across in their writing, where the author tends to position themselves as outside, or even transcending time and space, speaking universal truths down to the hoi polloi. We are trained to see, respect, and imitate that positionality because it is supposed to be "objective". Almost all of us write from that positionality, most often because we are forced to—if we want to be published, secure, and keep jobs and a salary.

To be sure, academe also allows for solipsism—where we take ourselves as our own body of knowledge. Yet, like the Enlightenment "transcendental" position, that attitude of omniscience remains—you can't question me because it's my own body, my own life. We need to move away from this rigid insistence on the infallibility of the author and acknowledge that life is just too damn complicated for any of us to be able to render wholly universal truths. I think intersectionality requires humility and rigor, dogged determination, and modesty. We can't stand on soapboxes and demand the rest of the world do as we say, and not as we do. I don't think that does much beyond convince ourselves of our own rectitude.

Q2: Assisted Reproductive Technologies, LGBTQIA+ Rights, and the New "Gender Trouble"

Tania Kupczak's chapter in *Domain Errors!* "So You Wanna Be a Cyborg Mommy? Queer Identity and New Reproductive Technologies" was devoted to queer parenting, pioneering in its discussion in relation to the early ARTs (Kupczak 2002). Today, Jaya Keaney's work, for example, on racialized intimacy, Indigeneity, ARTs, and queer parenting, contextualizes decisions surrounding agency, access, care, and responsibility (Keaney 2023). Thus, since *Domain Errors!* came out with its chapter on queer parenting, queer access to reproductive technologies has certainly impacted mainstream acceptance of IVF and surrogacy, and changed how male pregnancy, or trans- and gay parenting are approached as current issues.

The trans community's work renewed attention to questions of embodiment, definitions of sex, gender, and parenthood, and the role of technology, science, and biomedicine. The non-binary and LGBTQIA+ community itself is intersectional, and not all voices are equally heard or present. We are specifically thinking here about the work of trans- and queer artists and scholars who are highlighting topics in anti-blackness, settler colonialism, and environmental injustice. In parallel to them, we witness the return of (often divisive) feminist debates around the definitions of sex, gender, women, and men.

All this work takes place during the new moral panics around the globe, following a period of seemingly mainstream acceptance of queer rights and visibility, at least in some parts of the world. These panics and governmental support that legitimizes them lead to violent transphobia, reactionary anti-LBGTQIA+ legislation, book bans, and medical obstruction on a global scale.

What Do You Find the Most Helpful in This Current Moment of Research and Creative Work Around Sex, Gender, Sexuality, and Parenting?

Tania Kupczak

I wrote this essay as a grad student in 2001. Looking back on it now, I have many critiques of the scope of my writing, most notably my misperception that "queer" and "gays and lesbians" were interchangeable, as well as the now-glaring omission of nonbinary and trans folks. Language, which was one of the focuses of that essay, has been a great area of learning for me. In this moment, I'm very excited about the current trends in neopronouns and self-defined identity markers. The potential of language to dismantle binaries is profound, if our colonized minds can be opened to it!

María Fernández
The work of younger scholars, students, and artists, such as the ones mentioned in this conversation, is crucial to remind me that creative interventions continue regardless of repression.

Amelia Jones
Given the situation I identified above, the best thing we can do intellectually and creatively is to keep conversations going among us and to make these as public and as generative and generous as possible. Teaching classes with openness and directness (to expose these structures and generate debates about alternatives), running conferences, producing books, exhibitions, articles, films, and artworks that interrogate the virulent backlash strategies we find growing around the world, opens the door for forging coalitions and solidarities. The most dangerous moments are those in which these solidarities fall away and we become fragmented and disconnected, fighting amongst ourselves, fulfilling the fragmentation required by the mechanisms of late capitalism and late neoliberalism, which function best when we are divided and obsessively focused on our individual concerns and cares rather than connected to others. Even just providing an opportunity to identify and describe what is behind the backlash strategies is a political act in that it brings us together in recognition, giving a huge sense of relief (see below on my co-organized conference panel in Bristol).

This "virtual roundtable" is, precisely, an example of keeping conversations going among us so that we can refuse the alienation of twenty-first-century late capitalism.

Laleh Mehran
As a university professor, I have supported students through gender transitions over the decades, starting with the fundamental respect of using their correct pronouns. I also create space for students to explore sex, gender, sexuality, and family through project assignments that encourage personal expression and critical engagement. Many have examined "hacking

normativity" by challenging traditional family structures and reimagining caregiving beyond heteronormative and nuclear family models.

Art as a form of political resistance brings these complexities to life, offering new perspectives and possibilities. Conversations about gender, sexuality, and parenting must also consider race, migration, and ethnic knowledge systems that challenge narrow, state-sanctioned views of family norms. In a world where algorithmic systems reinforce biases—misgendering individuals or censoring queer content—feminist and queer data practices, archival projects, open-source databases, and alternative ways of organizing information are essential to ensuring nonbinary, trans, and queer stories are heard. Creative coding, DIY tech tools, and tactical media remain powerful strategies for resisting systemic oppression and driving change.

Susanna Paasonen

A huge and important question. There is a backlash for sure, and one that comes from various directions, or in multiple waves, so that there are obviously different kinds of helpful work to be read and done. LGBTQIA+ legislation has changed drastically for the better during the past decades, even with the current backlash, and anti-trans panics are inseparable from how trans and non-binary identifications have become acknowledged and publicly represented.

My own work for the past 20 years has been more on critical sexuality studies than gender, and here we've seen drastic shifts in the volume of research published, as well as in the topics explored. I have little to say about parenting or reproduction, really, as these are not my areas of work. But in sexuality studies, there's really crucial increase in the voices of sex workers in debates concerning platform labour, for example, as well as a drastic wave of porn scholarship addressing both historical and contemporary developments; queer theory pretty much weaves in and out of this all, together with intersectional analytical lenses (even as, sure, these are not always equally distributed). In the current moment of backlash, which extends to sexual rights broadly construed, scholarship engaging with advocacy work is truly crucial.

Nell Tenhaaf

I live in a very open culture and so I can only advocate for continued public language that is clear and repeated, e.g., I admire statements like this about Toronto artist Zavisha Chromicz: "As a self-identifying queer fat trans mixed

Roma self-taught artist who has been making community-based mixed media and fibre-based art for over twenty years, Chromicz views this practice as a medicine for survival, exploring disability in a throwaway culture and honouring survivors of childhood and ancestral trauma." (Chromicz 2025, from Toronto gallery Paul Petro Contemporary Art site, my gallerist).

Q3: Reproductive Labor/ Reproductive Justice

In 2022, the US Supreme Court overturned 1973's Roe v. Wade. In response, countries like France and some US States (e.g., Michigan) have moved to protect reproductive rights by constitutionalizing them.

Meanwhile, on February 16, 2024, the Alabama Supreme Court issued a ruling declaring that all embryos created as a result of ARTs, including IVF (In Vitro Fertilization), would be considered children. This means that any "embryo" (based on the legal definition - how many cells have been divided for how long, in a fertilized egg) that is no longer viable for uterine implantation and is discarded, should be given the status of a murdered person (a murdered minor). Currently, 500,000+ global births take place annually using IVF. With the normalization of surrogacy, including transnational and international surrogacy, its use has skyrocketed. Much of *Domain Errors!* second section—"The Female Flesh Commodities Lab"—was devoted to making visible the cost, physical and emotional labor, of those whose bodies are ground zero of these technologies and the desires that underpin them. Today, the work of Neda Atanasoski, Kalindi Vora, and Dána-Ain Davis, among others, brings new questions into these discussions.

What Do You Think of This Current Moment in Relation to Reproductive Labor, Justice, and Technoscience?

Tania Kupczak

Having been wrapped in labor organizing, mutual aid, and pro-abortion work for the last decade, the most salient message I hear over and over again is "We will save ourselves." Capitalist systems will never support the interests of marginalized communities. We must create our own systems of care, including a rejection of the absolutism of legality, in a framework of repro justice and mutual aid.

Embodied knowledge must keep rising as a resistance to patriarchal power dynamics in legislative, medical, and academic spheres. A great example is the Four Thieves Vinegar Collective, an anarchist biohacking group that offers recipes for DIY pharmaceuticals. The use of misoprostol in self-managed abortions owes a debt to a group of Brazilian feminists in the 1980s who began testing medications with pregnancy warnings. Here in the US, we can learn so much from our comrades around the world who have been resisting oligarchical regimes for generations.

María Fernández
We must think of reproductive politics as larger than biological reproduction. Laura Briggs' argument that all politics is reproductive politics is apt here. As she explains, immigration, living wages, education, housing, health, child and elder care, are all part of reproduction. To these we could add a healthy environment. Now there are more options to create families than ever before, from ARTs to family selection based on affinities. After Keaney, I suggest that we consider the consequences of our choices in broader social and political spheres. What effects do they have on our immediate families and in our communities? Who or what do they impact further away and how? Atanasoski and Vora's historical examination of surrogacy (2019) in relation to robotic technologies and race reminds us that technologies are never free from culture and invites us to think about people in various contexts who support reproduction in the wider sense. The early work of Saskia Sassen on labor, capital, and technology was prescient in this respect.

Amelia Jones
The right and far-right seem to be developing increasingly aggressive but also sophisticated strategies to control our bodies (particularly those of queers, trans, and women-identified people). If it weren't just baldly horrifying and dangerous, it would be intellectually and politically fascinating to watch the far right roll out strategy after strategy. Other than the epochally terrible overturning of Roe v. Wade, in the U.S. many of their machinations to date have, very fortunately, failed and we seem to be in a counter-backlash at least in terms of abortion rights, where they have lost a lot of credibility and votes by stacking the Supreme Court and pushing to eliminate the right to choose in the United States.

That said, we have to be super vigilant because their strategies are continually evolving. We need to agitate continually to get abortion rights enshrined constitutionally—at least in state constitutions—and to maintain the right to ship and supply abortifacients such as mifepristone across state and national borders (the work of Rebecca Gomperts and the Women on Waves initiative are extremely galvanizing in this respect). We need to articulate our message clearly so that even many on the other "side" agree with our insistence on the right to choose (depending on the survey, 55 to 70% of Americans believe in a woman's right to access full reproductive healthcare, including abortion). But we also need to continue to interrogate the class-based privilege of reproductive rights technologies as they are developed and applied (I pointed this out in my essay in the *Domain Errors!* book already).

My guess is that laws against IVF and other means of improving birth rates will fail for the wrong reasons. They will fail because most nation-states will be increasingly focused on maintaining the growth model of (late-) capitalism, even as the world burns—governments are already scrambling to "encourage" women to have more children (bizarre in the face of climate change, and only explainable through the fact that most countries are still basing their economies and societies on the capitalist growth model). That is, even states identified as "communist" (such as China) are now pushing for higher birth rates (versus the previous "one child policy") as they invest ever more dramatically in late capitalism as a means of increasing their global power. So laws lowering the chances of "growing" populations will be stifled. Sadly, this will contravene the only positive aspect of lower fertility: the decline of human reproduction viewed abstractly is a great thing for the health of our planet. Thus, the right policy (maintaining legal IVF processes) for the wrong reason (increasing populations).

Laleh Mehran
With increasing restrictions on abortion and gender-affirming care, underground networks providing self-managed reproductive health services are re-emerging. While essential for many, these networks can't fully replace comprehensive healthcare, leaving critical gaps in access. This underscores the deeper ties between reproductive justice, state control, surveillance, and climate collapse, raising urgent questions about whose bodies are controlled, exploited, or denied care.

I'm particularly inspired by Dr. Harriet A. Washington's (2008) *Medical Apartheid*, which examines racial inequities in healthcare and argues for community-led approaches that dismantle institutionalized biases rather than relying solely on policy reforms.

I'm also deeply concerned about how digital surveillance is tightening control over reproductive autonomy. Tracking menstrual cycles, location data, and online searches has become a tool of criminalization, making awareness and counter-surveillance strategies more urgent than ever.

Many apps collect and share sensitive data, often without users realizing the risks. I actively discuss safer alternatives, digital misinformation tactics, and data obfuscation tools with students and peers to help protect both ourselves and vulnerable communities.

At the same time, reproductive technologies like IVF are caught in political and legal battles over personhood laws and government control. While offering opportunities for parenthood, IVF is increasingly subject to legal and ethical restrictions. Surrogacy and egg donation industries often rely on lower-income women in countries like India, Ukraine, and Mexico, raising concerns about exploitation and power imbalances. Reproductive justice is also deeply linked to environmental justice—climate change, pollution, and heat stress disproportionately impact Black, Indigenous, and poor communities, exacerbating existing inequalities and further restricting access to care.

A powerful artwork that brings much of this to the surface is Lauren Lee McCarthy's *Surrogate* project, offering a deep and thought-provoking exploration of reproductive technology and bodily autonomy. This multifaceted body of work includes an interactive live performance, a one-woman show that immerses the audience in the experience of becoming a remote-controlled surrogate.

Susanna Paasonen

It's basically absurd, especially if we're looking at the US and how "pro-lifers" have issues with IVF. Yes, of course, there are huge ethical issues connected to surrogacy across gendered geopolitical inequalities that need to be attended to, but this is hardly a concern of theirs. This return to an imaginary organic realm of reproduction where no abortions may occur, where access to contraceptives isn't there, and where tech doesn't enter the picture is, well, absurd. I believe it's a common experience for women who have sex with men to make extensive effort throughout their lives not to get pregnant, so that reproductive sex is likely to happen only on very special occasions, if it happens at all. That is, when it's been especially planned.

So, this reproductive imaginary of freely flowing semen and receptive lady-bits fails to connect to the reality of checking ovulation rhythms, testing semen and hormones, and timing things that have little connection to either spontaneity or to how people regularly have sex. In this context, justice is much more concerned with the rights of embryos than those of actual children, or let alone their mothers (independent of their gender identifications), which is enough said about the ethics, or indeed morals, of it all.

Very dispiriting, but against the backlash, it's important to acknowledge the places where things have been moving forward in terms of reproductive rights (Ireland! Even Finland!) and where backlash has been met with serious pushback (Poland!). But then there's also the anti-trans, radical feminist activity with ties to the conservative right, which is attacking things like sex education curricula in the UK for being about "gender ideology", for normalizing queerness, and for being pro-trans. So, we have a situation where the World Health Organization's sex education and sexual rights recommendations are being attacked by vigilant feminists. Absurd! Not funny, but absurd.

Nell Tenhaaf

I haven't thought much about these issues apart from following in the news the tumult in the US around abortion and IVF. I live in a more liberal country where, so far, these are protected. My ethical position, and my belief since I was young, is that we don't need more children on the planet than we can fully care for, and in parallel, that overpopulation of the planet leading to degradation of the planet itself is a crisis now overwhelming humanity. Women everywhere are on the front lines of this dilemma, but solutions have to come from the social body—no one can bear them as an individual.

In the 1980s, my work was suffused with feminist issues around biotechnological invasion of the body space and the control, hegemony, and ethos of the 'master molecule' (DNA) surrounding developments in genetics. That seems like a long time ago, and I pivoted from such an overt critique of the sciences to find strategies for aligning with more relational sciences like the 'social intelligence' branch of AI. For a time, I was questioned about what happened to the feminist politics in my work—but it still infuses everything I do.

Q4: Art Strategies and Solidarity Tactics

Artists, designers, IT professionals, and activists experimented with many media and aesthetic strategies in the 1990s. subRosa was founded in the early days of the World Wide Web, along with groups like the OBN, ®™ark,

The Yes Men, and Critical Art Ensemble that employed the aesthetic strategies of satire and spoofing, among others. Other bioartists working with tissue culture, such as SymbioticA, Eduardo Kac, Stelarc, Paul Vanouse, and Adam Zaretsky, also used bio-hacktivism.

At the same time, Tactical- and Indy-Media used strategies of police counter-surveillance, culture jamming, open access, and self-publishing from citizen journalists, activists, and art movements. subRosa members knew or collaborated with many of these artists.

subRosa, as a cyberfeminist art collective, used humor and satire, too, however, it situated itself within the legacy of the Feminist Art Movement and Feminist Technoscience, and gathered combinations of strategies that were also informed by Marxist and anarchist critique, labor, and anti-war movements.

Domain Errors! reflects that. subRosa worked with an independent publisher, Autonomedia, that sees itself as "an autonomous zone for arts radicals in both old and new media." The book included an anti-copyright statement. The third section—"Research! Action! Embodiment! Conviviality!"—included a list of "subRosa Manifestations" and discussions of feminist art in Singapore, empathetic cybernetics, and the menopausal cyborg. This third section was based on feminist solidarity tactics.

Today, feminist art, activist, and technoscience communities develop strategies around joy, killjoy, care studies, crip technoscience, sustainability, and speculative and imaginative work of change, in addition to humor, subversive critique, and many other strategies and tactics. Octavia Butler and Silvia Wynter have become important figures in this work. Our values stay (acknowledgment of reproductive labor, justice work, non-violent activism, art, and scholarship as tactics for change), but our art and scholarship tactics and methods have evolved.

What Art and Solidarity Tactics Inspire You the Most Right Now in Your Scholarship, Creative Work, and/or Teaching?

Tania Kupczak

As an artist entirely outside of academia these days, I draw a lot of inspiration from other activists and grassroots groups that use creativity as a message-delivery method. The folks at Beautiful Trouble have great resources for public actions. I collect zines and micro-publications from artists, such as Daisy Salinas of *Muchacha Fanzine*. The immediacy and unapologetic nature of this art form push me to spend more energy on community engagement and mutual aid rather than trying to make a perfect object. I love the swell of encouragement coming out of youth movements that anyone with a story to tell or an idea to share can find a creative way to do it.

María Fernández
Violence, repressive regimes, governmental ineptitude, and sometimes indifference have led many Latin American artists to address social injustices in their work. At the same time that artists' engagement with technologies has blossomed, there has been increased attention among them to racial, indigenous, and ecological activism, feminist, and LGBTQ rights. This work is generative, as traditionally, some of these subjects were underrepresented in the history of art of the region. It has led to expansions and revisions of the canon, which affects the way we think about Latin America, modernity, and feminism.

In the social sphere, I find interventions that are performative rather than declarative most valuable. These do not have to be grand but can be as modest as gestures of support and acts of kindness for others in our communities and beyond. Especially after the emergence of #MeToo, Black Lives Matter, anticolonial, indigenous, and social movements across the globe, it has become an unspoken requirement in certain spheres to assert alliance to feminist, antiracist, and decolonial politics as a demonstration of progressive credentials. Words alone sometimes result in inertia.

Amelia Jones
I'm totally claiming your "menopausal cyborg" label! (Not the least because I have a titanium hip and just had my ovaries, riddled with giant tumors, removed.)

Again, I would point to the groups insisting on forming coalitions, spreading the word, generating awareness and debate, and crystallizing tactics. Exemplary collectives include subRosa, RAQs Media Collective (New Delhi), or, in Los Angeles, Lauren Bon's collaborative Metabolic Studios *Bending the River* project, which is based on a decolonizing approach to revitalizing the LA River through creative practice, attending in a feminist way to our closeness to and dependence on the well being of the land and a decolonial attention to the original human stewards of the LA River basin—the Tongva and related Indigenous tribes. Bon is a person with resources who is also an artistic visionary who brings together large networks of creative people to have conversations and take action—she has, along with organizations like Clockshop, reframed conversations and reshaped municipal policy around the crucial role of the river in sustaining life in the Los Angeles basin. Among other initiatives, including redirecting the LA River to plant a cornfield in a then-derelict tract of land next to the river, she has co-taught a class for my school (USC Roski School of Art and Design) with my Roski colleague and feminist activist artist Suzanne Lacy; the class brought students into the center of *Bending the River*, and featured as a performance studies class for the initiative I discuss below.

In a different community in Los Angeles, Crenshaw Dairy Mart, an art collective co-founded by three former USC Roski MFA students—Patrisse Cullors (also co-founder of Black Lives Matter), noé olivos, and ali reza dorriz—works to create communal conversations and arts events to connect the power of the arts to the existing creative cultures of South Central Los Angeles. Again, coalitional conversations and creative productions (such as their abolitionist pod—a mobile garden that they bring into low-income neighborhoods that lack greenery and fresh produce as well as art venues) are elaborated using low-tech means to change the power dynamics of the late capitalist metropolis.

The most fantastic feminist queer/trans cyborg work I follow is that of Young Joon Kwak. I just two days ago witnessed a performance of hers in the context of an LA Freewaves event, Femmes, held in Los Angeles Historic Park. Kwak had three young queer bodies intertwine tightly, covered them with a gloopy silicon-like substance, and then cast the resulting mold into a sculpture. Kwak has an unerring eye towards subverting public and gallery spaces, hijacking them for her radical queer agenda. She uses bodies and modes of glittery sculptural installation to change the (gendered) meaning of spaces.

Finally, I will note my own current project, interrogating and putting pressure on the structures of neoliberalism that have compromised institutional life, including at universities and across the art world. I am working on a single-authored book (entitled *Lifework: Against Cultural Capitalism*), which critically examines the current functioning (or dysfunctioning) of the university and art world through specific case studies, but also looks at artists and writers/curators who have lived and created successfully against the grain of capital over the past 50 years. But more to the point of your question I'm also working collectively via mounting a series of events/panels/workshops at various conferences across the US/UK (so far; we hope to go wider) with my colleague Dr. Ben Nicholson to generate more discussion and to activate the kind of recognition

I noted above as crucial to solidarity and to countering fragmentation. These bring people together to articulate the difficulties of working in neoliberal universities and dealing with the blatant contradictions of the supposedly liberal art world in the twenty-first century. For example, the panel Ben and I organized at the Association of Art Historians at Bristol University in the UK (entitled "Selling Out? The Neoliberalism of the Art World and Academia") allowed for a crowded room full of participants and audience members to join in an extended conversation around our varied experiences; people were moved almost to tears at finally having their anxieties and exhaustion recognized as system-caused rather than as individual failure, as we are generally taught to feel. Notably, a similar panel proposal for the US version of the AAH—the College Art Association annual conference—was summarily rejected, but we are trying again this year, and will bring a similar initiative, this time in workshop form, to the Association for the Study of Arts of the Present (ASAP) in New York in October. These conversations will result in an anthology, entitled provisionally *System Failure: Surviving Academia and the Art World in the Twenty-first Century*. But, more importantly, they ratify our experience (almost universally horrible) working in institutions that brand themselves through high-minded rhetoric of research and creativity but blatantly belie and even work against these ideals through their not-so hidden agendas, which focus on branding, "efficiency" (not only impossible but not desirable for any institution of teaching, learning, research, and creativity), and rely on the federal government or networks of extremely wealthy people, mostly from the corporate realm, to fund and direct them. By bringing people together, again, we refuse the fragmentation that these institutions implement in order to keep their hidden agendas in force. We can strategize and push against the counter-productive and toxic changes higher education and arts organizations have implemented since the rise of neoliberalism around 1980.

Laleh Mehran

To uphold solidarity and intersectionality, I dedicate a great deal of labor across various initiatives. As the Associate Director of the Clinic for Open Source Arts (COSA), I help drive our mission to support new open-source projects in the arts through mini-grants, residencies, workshops, and training and assistance to a diverse set of voices in making tools more accessible. We focus on fostering and growing projects, particularly those that amplify underrepresented perspectives.

I also dedicate significant effort to supporting underrepresented groups within my academic institutions. I've served on the Executive Board of Gender, Women's, and Sexuality Studies, mentor to new faculty and first-generation students, and, alongside two colleagues, helped establish an affinity group for faculty who identify as women, trans, non-binary, or gender- nonconforming people of color. Through this group, we exposed the lack of transparency and inconsistent procedures across disciplines and divisions, issues that were brought to light and, in some cases, addressed. While these roles are formally recognized by the university, the immense invisible labor and emotional support required in these spaces remain an ongoing and heavy burden, one that disproportionately falls on women of color.

Beyond academia, I serve on multiple nonprofit cultural boards, mentor alumni, and organize numerous events, including Momentum: Women/Art/Technology (with support from Muriel Magenta, highlighting the intersection of women's contributions to art and technology). I work to foster community and dismantle gatekeeping by creating spaces where people feel empowered to take up space, share knowledge, and build networks of support that challenge systemic barriers.

With political uncertainty and increasing hostility, coming together in person has never felt more crucial. As a professor and mentor, I often organize in-person events to connect creatives from different communities, strengthening solidarity networks and fostering safety in numbers.

Physical presence builds a sense of connection that digital spaces, despite their reach, can't fully replicate. Whether in classrooms, community spaces, public demonstrations, or intimate gatherings, coming together generates energy that builds trust and reaffirms shared values. Face-to-face interactions enable deeper conversations, collective strategizing, and emotional support, things that are harder to sustain in a digital landscape shaped by surveillance, censorship, and corporate control. In these spaces, I'm reminded that I'm not alone in these struggles. I've seen how in-person gatherings create opportunities for spontaneous dialogue, shared meals, and moments of care that are vital to resistance. In politically difficult times, when marginalized communities face constant threats, these real-world connections sustain movements, nurture resilience, and reinforce the idea that solidarity isn't just theoretical, it's lived and strengthened when we show up for each other.

Educational practices play a crucial role in reshaping these conversations. When I am teaching AI in creative projects, we examine how even basic prompts can generate biased results. As students uncover these biases, they become more critical of auto-generated outcomes. By exploring the spread of misinformation, whether through deep fakes, algorithmic propaganda, or counter-mapping exercises, students develop skills to critically navigate digital spaces.

Speculative fiction and worldbuilding offer ways to reimagine history and the future beyond oppressive paradigms, while augmented reality reveals hidden histories of resistance.

Experimenting with avatars, deepface drag, and virtual embodiment can serve as solidarity tactics. Many students are increasingly comfortable using these tools to assert their identities and connect with others in empowering, creative ways.

I often share with my students the metaphor of an onion to illustrate solidarity and intersectionality. Just as an onion has layers, solidarity involves both giving and receiving protection. While I can often shield my students, providing them with space and support to navigate their own layers, I, too, need protection in different ways, whether through peers, institutions, or collective action. At the same time, many students don't realize they have the power to support one another, especially those in safer situations who can step up for peers facing greater risks. I encourage them to recognize their own agency and take action, fostering connections that create a sense of safety, belonging, and support for everyone.

Grassroots movements outside traditional institutions are reshaping knowledge-sharing and sustainability by challenging capitalist education models and providing support to marginalized creators through community hubs such as hacker/maker spaces. When it comes to creating art spaces that are sustainable and non-exploitative, we have to make sure social justice is deeply embedded, not just a concept. I've learned from student-led solidarity movements, especially those organizing outside of corporate social media, about the hidden costs of "free" apps and how data and content creation often fuel exploitative digital economies. In my courses, I focus on fostering both critical thinking and compassion, merging creative work with resistance and collective empowerment. Building trust with students takes time, but once that trust is established, they often share that they gravitate toward me because they believe I may have a lived experience that resonates with their own, helping them feel more connected and understood. Creating a space where students can bring their personal experiences and beliefs into their art empowers them to create impactful work.

Susanna Paasonen

I have definitely been thinking a lot about joy and excitement, both within and outside the framework of affect inquiry, for quite a while now. Together with Jenny Sundén, we did a book in 2020 on feminist social media tactics within/in the aftermath of #metoo where we argued for the importance of laughter—and, indeed, for the potentials of absurdity, which have nothing to do with my point on absurdity just above—in how activism is sustained and in how people keep going among all the hate and bile that network cultures currently spew and foster. Anger, rage, and sorrow are all efficient in pushing individual and collective bodies into action, but negative effects also eat away at people's capacities to act; it wears us down so that the figure of the killjoy isn't quite enough when it comes to resistant tactics.

Pockets of joy, no matter how fleeting they are or how frivolous they may seem, and no matter how out of place or inappropriate giggles may feel at any given moment, these are things to hold on to and to value. Laughing together is valuable, even as it's important to acknowledge that we do not need to be moved or amused by the same things to do politics together. We do not need to feel similarly in order to have the same goals.

To rephrase, absurdity can pack a lot of punch as a resistant tactic by virtue of refusing the logic on offer ("this is ridiculous!"), by turning things ridiculous through tactics like exaggeration, literal reading, or parody, and by turning things upside down, or inside out, to allow for a different place to start through interruption. This logic (to the degree that it is one) was key to the culture jamming activities two some decades ago, and I truly think it's something to hold on to.

Nell Tenhaaf

I am still interested in reconciling the 'coldness and care' exemplified in the lab coat, which I spoke about 20 years ago in *Domain Errors!*. But now I think about it as how to direct our help, related to bridging the nature/technology gap, i.e., humans need help, nature's entities need help, and our tech artifacts need help if they are to be related to us – the politics of mutual help, both rational and emotional.

In the current reworking of older electronic works from the early and mid-2000s, I'm placing clearer emphasis on the social relations aspect of AI, or how we must learn to deal with the power dynamic between us and the new 'them'. In fact, I prefer the concept of 'artificial agency' to AI, which, as a cultural meme, has now come to signal, on the one hand, a fear of being overpowered, and on the other, an expectation of being rescued from ourselves. My zone for several decades has been artificial agency that functions apart from human mimicry, i.e., artifacts that have their own innately machinic forms and expressions. The social bonds and dynamics (of control, caring, attachment) that we humans form with these 'bodies' are at the core of what I'm working with. All of this evolved out of the feminist critique of scientific work that preceded it.

Further on the topic of revisiting ideas in older works, with a view to the path forward needing to also have a path backward (as I talk about above), I am always interested in reconciliation strategies for settler artists and writers to forge solidarity with historically marginalized people and groups, especially Indigenous. Some recent innovative writing that reconsiders the past breathes new life into theoretical alliances between now well-established feminist thought and Indigenous/racialized/queer arts activism. An example that is personal for me is an article that probes an innovative and early (1989) exhibition of electronic and computational art in Canada, where the author looks at an 1988 electronic media work of mine that focused on gender and class issues. She discusses this work, *Gramatica*, with a view to the fallout from modernity's exclusionist tactics, and gives it new life in parallel with her own settler solidarity with Indigenous peoples. How rewarding is that! (This is Jessica Jacobson-Konefall's "Beyond Electronics: Pimatsiwin and Computing Arts in Manitoba" in Adam Lauder's *Variable Conditions: Paracomputational Arts in Canada*, 2023. 'Pimatsiwin' is a Cree expression that means 'a life of learning and growth').

Michelle M. Wright

I just taught an undergraduate literature course on LGBTQ African American literature, and I was taken aback at how critical all my gay, lesbian and straight Black students—except the one and only trans student—were of the Black lesbian readings, in this case Ann Allen Shockley's *Loving Her* (2014/1974, racist), Audre Lorde's *Zami* (1982, irrelevant) and Jewelle Gomez' *The Gilda Stories* (1991, coercive vampirism). Appositely, they loved our Black gay male authors—James Baldwin (*Giovanni's Room*, 1956), Randall Kenan (A Visitation of Spirits, 1989) and Brandon Taylor (*Real Life*, 2020), but in a way that I saw as very rigid, meaning they imposed their own racial politics on the text, and didn't want to discuss the many ways the texts failed to cohere to those politics. I was also thrown by how few Black students took the course—it was overwhelmingly white women, and barely a third of the class were people of color, which is unusual for me at Emory.

So, I am still mulling over that course, which, from a pedagogical point of view, was very successful because we had very full, passionate but respectful discussions, and all of the students developed their analytical skills and wrote sharp, insightful papers. I think the changes I am witnessing are largely in how students seem to be asking much more searching, detailed questions than earlier generations, and the range of questions and ideas they had was equally astonishing— everything from food and drinking habits, to urban planning, to class tensions (i.e., between the middle class and the working class). I can only project on to them, of course, I don't know their minds, but my sense is we shared a dread of the current political moment, and agree that we are all truly in a dangerous place because the desire to control, exploit, harm and eradicate difference has moved from the fringe and cloistered communities right into the public square in broad daylight. Unlike my future as an undergraduate in the Reagan and Bush years, their future looks not only politically bleak, but economically and ecologically devastating. So I think they are looking to literature for allyship and political guidance, and have less time for what they see as abstraction, ambivalence, and ambiguity. They want clarity, and feel safer with Black male expression and viewpoints, which is crushing to me because it hurts to hear a brilliant strategist and thinker like Audre Lorde dismissed because they find her interest in the erotic as a site for individual agency and expression at best irrelevant, at worst perverted. As more than one person put it, narratives of despair strike them as more honest than narratives of hope and resistance.

Now, of course, this is from those students who spoke several times every class session, versus the quieter ones who spoke only once or twice per session. But honestly? I wonder what I am missing, and what my age and own rigid opinions make me deaf, dumb, and blind to, even when it's being expressed to me directly in the classroom.

Q5: Learning from Our Students—Teaching Topics from *Domain Errors!*

The editors and authors of *Domain Errors!* were artists, educators, and graduate students. Since its publication in 2002, higher education and students have changed a lot. Academic positions in the arts and humanities, and tenure, are under attack. Our students are participating in the new progressive or conservative movements, often referencing the social movements of the 1970s and the 1980s. At the same time, most current undergraduate students have never experienced the world before 9/11, social media, cryptocurrency, YouTube, Google, AI, and smartphones.

Presumably, a greater percentage of our students are now born as the result of reproductive technologies, to which at the time we felt a combination of hope, fear, and ambivalence, as reflected in *Domain Errors!*.

How Has Your Teaching of Cyberfeminism, Technoscience, or Sex- and Gender-Related Topics Evolved in Relation to Changes in Your Students, Your Local Context, or Your Own Thinking/Making? What Are You Teaching Right Now?

María Fernández
Some years ago, I began working on cybernetics. Through that work, I became interested in nonhuman agencies and in the history of feminist art that engaged with them. In the fields I teach, histories are important to me because it is so easy for us and our students to focus exclusively on the present and forget the achievements and challenges of past generations. Currently, I teach a lecture course in the history of feminist art in new media, a seminar on posthumanist feminisms, and, prompted by Mindy Seu's book, I resumed teaching a course on cyberfeminism, which I had not taught for several years.

Amelia Jones
I teach now in an art school within a larger university. I was brought to USC Roski School of Art and Design in 2014 to work with colleagues to completely rebuild the critical studies curriculum, and so I was able to rehaul the outdated curriculum with my colleagues. We developed a carefully thought-through series of classes that (on both the undergraduate and graduate levels) build from a basic understanding of the history of the invention of the concept of "art" and of the arts complex (including

higher education arts training as well as museums, galleries, etc.) to instill a critical sensibility in our students. They come out of our programs (and my classes) with a pretty incisive sense of the power structures and ideologies at work in the various potential arts and curatorial worlds they will enter into, including a deep understanding of the history of colonialism and extraction since they played such a central role in the European invention of "art" as we know it. I have also developed a graduate certificate in Performance Studies, which works across programs at USC to make use of stellar faculty from Dramatic Arts, Dance, and other arts and humanities areas to provide an understanding of the field and practices connected to the discipline of performance studies. My core course for this certificate teaches core texts in the field from its establishment beginning in the post WWII period to the radical recent contributions by Black Studies feminist performance scholars such as Daphne Brooks and Zakiyyah Iman Jackson (who was USC faculty until recently and has visited my class), both of whom address histories of technologies of representation/presentation in music and art to understand more fully the discursive production and bodily experience of Blackness, queerness, and femininity as these intersectional identifications condition broader American and Western culture as a whole. This is the initiative for which Lauren Bon and Suzanne Lacy taught their maverick class through which they immersed students in the *Bending the River* project.

Laleh Mehran
Today's students have grown up immersed in social media, AI, and algorithmic influence, shaping their intuitive understanding of digital surveillance, platform labor, and gender representation in online spaces. Unlike previous generations, they are not just witnessing the digital transformation of gender and labor; they are living it. At the same time, many students are grappling with digital oppression, from AI misgendering and cyberbullying to the shadow-banning of marginalized content. Some actively resist these systems, while others feel overwhelmed by the scale of systemic issues. In addition, many students express deep concerns over climate collapse, political instability, and economic precarity. Yet, there is also a rise in apathy, possibly due to a sense of powerlessness in the face of forces beyond their control. This shift may also explain why privacy concerns have diminished; where earlier generations fought to protect their data, today's students often give it away freely on social media, resigned to the trade-offs of digital convenience. Physical safety remains an urgent

concern, particularly with the rising violent attacks on marginalized populations, as university campuses are increasingly safe spaces. Although these are seemingly insurmountable obstacles, I actively focus on equipping students with practical tools to address these major concerns. Methodologies include critically analyzing machine learning algorithms and their racial and gender biases, exploring speculative resistance through creative projects like worldbuilding. Through hands-on projects and collaborative work, students learn to apply these concepts in their own art and technology practices, empowering them to create more inclusive and equitable environments.

While students engage with contemporary digital economies, many are drawn to monetization strategies that reinforce exploitative systems. Some students have shared with me their aspirations to become influencers, a goal that I attempt to unpack critically, though I don't believe I have made a dissuading impact. This underscores the importance of interrogating who benefits from digital economies and who is erased.

Over the past two decades, I've observed a growing embrace of nonbinary, trans, decolonial, and posthuman perspectives. While not all students respect these positionalities, my experience across US higher education suggests a broader acceptance of diverse gender identities. I've also observed a significant shift in attitudes toward discrimination (though less so regarding race), where once students may have passively ignored disparaging remarks, many now openly challenge them, articulating why such comments are problematic.

Susanna Paasonen

These are people who've basically had smartphones all their lives, and for whom it's very hard, or even nonsensical, to think about everyday life as detached from network connectivity. So, this puts us in a very different place in terms of thinking about networked media since it's unlikely that discussions on feminism, or activism more broadly, or anything much, can be detached from it. I've increasingly been thinking of network connectivity as infrastructural to how everyday lives work, which then means that while it's crucial to question the political economy of it all (as in data capitalism, communicative capitalism, cognitive capitalism, or any other diagnosis of the conditions), it's equally crucial to attend to the nitty-gritty stuff of everyday life – how datafied lives are navigated and negotiated, the different kinds of vulnerabilities that emerge in connection with networked agencies, and what people do with it all.

Thinking of my contribution to the book, which came out from my doctoral work, so much has happened in terms of internet research as a field which, with the emergence of social media, saw a drastic influx of female and queer scholarship (while remaining, for a considerable while, pretty much white) and, of course, in terms of the very objects of study. Personally, I've moved from cultural critique of the kind that I learnt in 1990s academia to forms of investigation concerned with ambiguity – not as some limp sort of relativism, but as ways of acknowledging that things are complex; that mutually conflicting developments coexist and need to be equally attended to for the work to be critical in the sense of pertinence.

I teach media culture, both historically and with a focus on network culture, and occasionally do classes more specifically focused on sexual media and pornography. And I supervise a fair amount of PhD work at the intersections of sexuality, online stuff, and affect.

Q6: We Would Like to Hear about Your Current Interests

At the time when *Domain Errors!* was being written, subRosa and other cyberfeminist artists and scholars were inspired by—among other key texts—Donna Haraway's *A Cyborg Manifesto* and Sadie Plant's (1997) *Zeros + Ones*. Today, Legacy Russell's *Glitch Feminism* (2020) and Mindy Seu's *Cyberfeminism Index* (2023) are common touchpoints for those who are interested in cyberfeminism.

Are There Key Texts or Works of Art That Have Significantly Reshaped Your Work and Thinking in the 20+ Years That Followed Domain Errors!*? What Are You Currently Reading or Researching, and What Is the Most Interesting to You?*

Tania Kupczak

I'm such a discourse surfer that it would be hard to summarize the last 20+ years of exploration! I will always find beauty and solace in Felix Gonzales-Torres' work. When thinking about the evolution of cyberfeminism, I return again and again to my heavily-notated collection of CAE books, Afro-futuristic sci-fi novels, and the library of references I've gathered on self-managed abortion. Sage-Femme Collective's *Natural Liberty* (2008) is a great resource for reproductive bodily autonomy.

My relationship to technology has gotten much more complicated since *Domain Errors!* was published, and I now work with a broader definition to cyberfeminism that encompasses somatic knowledge and the pathways the natural world has been using for eons. We will save us <3

María Fernández
I am still working on cybernetics and learning about feminist posthumanisms. I find contemporary work in bioart particularly exciting. subRosa was concerned with biotechnologies since its foundation and I am grateful for its lasting influence on my intellectual development.

You already mentioned several important thinkers including Donna Haraway, Octavia Butler and Silvia Wynter. In addition, the work of Karen Barad, Elizabeth Grosz, Rosi Braidotti, Mel Chen, Hortense Spillers, Anna Tsing, Maria Puig de la Bellacasa, Lynn Margulis, Isabel Stengers, Wendy Chun, Silvia Rivera Cusicanqui, Astrida Neimanis, and Zakiyyah Iman Jackson, among others has been thought provoking for me.

Amelia Jones
As noted above, my book *Seeing Differently* immersed me in a more intersectional mode of thinking/being. My book *In Between Subjects: A Critical Genealogy of Queer Performance* (2020) was built on my increasing immersion in queer performance communities to examine critically the history of queer and performance discourses from the 1950s to the present moment, at which they have become mainstream. That is also related to an exhibition I curated that same year at Participant Inc. in New York, which travelled to the ICA LA, *Queer Communion: Ron Athey*. These latter two projects indicate the expansion of my feminism and its intersectionality. I still focus on embodiment and the politics of identification, but queer and trans- performance is a key interest (including the work of Zackary Drucker, Cassils, and Julie Tolentino, which I explore in *In Between Subjects*).

For the projects I am working on now, noted above, I am currently reading a lot of texts on neoliberalism, the late-capitalist art world, and the twenty-first-century university for the interrelated projects noted below. Naomi Klein's *Doppleganger* (2023) really stands out for me in this reading, not the least because it's actually fun to read (which can't be said of a lot of the texts on neoliberalism!). Black feminist theory and history—the books of authors such as Saiydiya Hartman, Zakiyyah

Iman Jackson, Sylvia Wynter, and Christina Sharpe—also guide my work. And I am reopening my fascination with French philosophy and identity theory by reading the excellent new biography of Frantz Fanon (by Adam Shatz 2024). As well I am curating a major survey exhibition of the work of Ken Gonzales-Day and so reading and editing the brilliant scholarly and creative essays that will be in the catalogue for that show, *Ken Gonzales-Day: History's "Nevermade."* This brings me back into post/decolonial theory, queer theory (Ken is a queer Latinx LA-based photographic artist whose work directly builds on his experience), and photo theory.

Laleh Mehran
There are numerous sources of inspiration; here are a few (in no particular order):
Labor & Gendered Work

- Laurel Ptak – *Wages for Facebook*, 2014. A reworking of the 1970s Wages for Housework movement, critiquing unpaid digital labor on social media.
- Sophie Lewis – *Full Surrogacy Now*, 2019. Radical argument for abolishing the nuclear family and reframing all labor as work.
- Tithi Bhattacharya, Cinzia Arruzza, Nancy Fraser – *Feminism for the 99%: A Manifesto*, 2019. A transnational feminist critique of neoliberal capitalism and how platform economies exploit gendered labor.

Bodies & Control

- Loretta J. Ross & Rickie Solinger – Reproductive Justice: *An Introduction*, 2017. Extends cyberfeminist critiques into the domain of reproductive surveillance and inequality.
- Gaia Bernstein – *Unwired: Gaining Control over Addictive Technologies*, 2023. Examines digital addiction and its impact on bodies, labor, and reproductive autonomy.
- Harriet A. Washington – *Medical Apartheid*, 2007. Exposes the long history of medical experimentation and systemic abuse of Black Americans in the U.S. healthcare system, highlighting how racism has shaped medical research and treatment.

Decolonial & Posthuman Feminism

- Kathryn Yusoff – *A Billion Black Anthropocenes or None*, 2018. Critiques the whiteness of geology and climate discourse, linking planetary destruction to racial capitalism.
- Kanta Dihal, Stephen Cave, Sarah Dillon – *Decolonizing Artificial Intelligence*, 2020. Investigates AI's colonial entanglements and ways to resist its extractive logic.

Global AI, Data Colonialism & Surveillance

- Safiya Umoja Noble – *Algorithms of Oppression*, 2018. Examines racial and gender bias in search engines and AI.
- Yanis Varoufakis – *Technofeudalism: What Killed Capitalism*, 2023. Frames digital platforms as extractive feudal systems.
- Nick Couldry, Ulises Mejías – *The Costs of Connection: How Data Is Colonizing Human Life and Appropriating It for Capitalism*, 2019. Argues that Big Data functions as a new colonial extractive system, dispossessing people in the Global South.
- Kate Crawford (2022) – *Atlas of AI: Power, Politics, and the Planetary Costs of Artificial Intelligence*, 2021. Maps the material and environmental toll of AI, including labor exploitation in global supply chains.
- Stephanie Dinkins – *Not The Only One (N'TOO)*, 2018. A conversational AI built from oral histories of a Black family, challenging racial bias in AI.
- Zach Blas – *Facial Weaponization Suite*, 2011–2014. Uses biometric distortions to resist facial recognition surveillance.
- Heather Dewey-Hagborg - Biohacker and artist critically examines technology, surveillance, and genetic privacy, questioning how DNA and biometric data are used to police, categorize, and control individuals.
- Mary Maggic – *Open Source Estrogen*, 2015–ongoing. DIY hormone hacking and bio-political interventions in gender.
- Dasha Ilina – *Center for Technological Pain*, 2019. Explores the physical consequences of digital labor and reproductive surveillance.

Queer & Trans Digital Bodies

- Danielle Brathwaite-Shirley and Black Trans Archive – *We Are Here Because of Those That Are Not*, 2020. Interactive archive centering Black trans lives in digital history.
- Sondra Perry – *Graft and Ash for a Three Monitor Workstation*, 2016. Explores racialized digital labor and avatar creation.
- Afsaneh Najmabadi – *Professing Selves: Transsexuality and Same-Sex Desire in Contemporary Iran*, 2013. Explores digital self-fashioning among trans communities in Iran under state surveillance.
- Afsaneh Najmabadi – *Women with Mustaches and Men without Beards*, 2005. Explores historical gender fluidity in Iran and its erasure through state control.

Decolonial Art & Race & Gender

- Neda Atanasoski & Kalindi Vora – Surrogate Humanity: *Race, Robots, and the Politics of Technological Futures*, 2019. A global feminist critique of automation, applicable to Iranian and Middle Eastern tech labor.
- Wafaa Bilal – *Domestic Tension*. In Domestic Tension, he lived in a gallery and was shot with paintballs by online users. Interactive artworks that explore themes of conflict, identity, and cultural politics, often reflecting his experiences as an Iraqi-American.
- Morehshin Allahyari – *She Who Sees the Unknown*, 2016–2020. Uses 3D scanning and digital storytelling to reconstruct erased Middle Eastern feminist mythologies.
- Suzanne Treister – *HEXEN 2.0*, 2009–2011. A visual history of cybernetic ideology and its ties to surveillance capitalism.
- American Artist: They legally changed their name to American Artist in 2013, to recontextualize the definition of the term "American artist." Their work focuses on themes surrounding "blackness, being, and resistance in the context of networked virtual life."
- Heba Y. Amin – *Operation Sunken Sea*, 2018–2022. A speculative counter-history of colonial water projects, critiquing technoutopianism.
- Raheleh Filsoofi – *Echoes of Silence*, 2021. Interactive installation using sound and biometric data to explore digital border surveillance.

- Mariam Ghani – *What We Left Unfinished*, 2019. Documentary film uses found footage and AI reconstruction to explore gender and political history in Afghanistan.
- The #WomanLifeFreedom (زن،یگدزن،یدازآ) movement, which gained global attention after the killing of Jina (Mahsa) Amini in 2022, has sparked a radical intersection of cyberfeminism, digital resistance, technoscience, and reproductive justice in Iran. Artists, activists, and scholars have used AI, data obfuscation, encryption, and digital storytelling to fight back against state surveillance, gender apartheid, and biopolitical control over women's bodies.

Susanna Paasonen

I've since engaged much with Wendy Chun's, Eve Kosofsky Sedgwick's, and Lauren Berlant's work in particular, and dived deeper into effect inquiry where Sara Ahmed was doing very interesting stuff already two decades + ago. From there, I've gone to Tomkins and even Spinoza.

So, that's sort of where I went; towards intersections of affect inquiry and network media, and also into queer-feminist sexuality studies, as I realized that online porn was so very under-studied. A large bulk of my work in the past 20 years has been on online porn, or online sexual cultures more broadly, but there's also the affect strand connected to media theory, and sometimes media history (I love archives). I read pretty much eclectically, as I believe that new avenues for thought come from looking at different stuff in parallel to one another, and in combining things in order to push forward to a different place.

One of the greatest joys of my academic career to date has been realizing the extent to which my stuff has run parallel to that of my Swedish twin Jenny Sundén, whom I first met in 2000, even as it took us several years to recognize our parallel tracks in cyberfeminist inquiry, and beyond. It's wonderful to have a close collaborator who's read the same stuff when it came out, and who's had time to think with the said stuff for quite a while now. Best things in academia, really, are these friendships that take shape, and matter, and that allow for thinking together.

And I doubt I'd ever contributed to *Domain Errors!* without being friends with María [Fernández], who's been my huge inspiration since we met when I was barely out of my teens (I was!). For me, what makes things worthwhile professionally are collabs based on friendship that come from a special place of trust and respect—love, basically—as these are the shelter that we need from all the politics going on inside and outside of academia. It's the stuff that keeps us going; the essential stuff of life.

Nell Tenhaaf
One of my favourite reading topics is digging into the history of computing, where I am still trying to wrap my head around how such a huge presence in our world arose mainly from the minds of geeky guys – are women stuck with this, or is there a hidden narrative to be dug out as there as, for example, in biological science (the resurrecting of Rosalind Franklin)? There's Ada Lovelace, of course, now at least often invoked in history. But what about the 20th century, when all of the core developments happened? In Walter Isaacson's *The Innovators: how a group of hackers, geniuses and geeks created the digital revolution* (2014), the central theme is collaboration at the core of computational tech development. Along with Lovelace, Isaacson calls attention to forgotten women like those who programmed ENIAC in 1946, Grace Hopper who wrote the first compiler. At least he's conscious of this problem. Then there's the marvelous actress Hedy Lamar who essentially invented Wifi, Bluetooth and GPS, she is sometimes profiled on the classic movie channel as an important inventor. In Jane Smiley *The Man Who Invented the Computer* (2010) there are a lot of mathematically-inclined wives standing by their men, a sign of the times not only in the 1940s but well beyond. And for a philosophical stance, in Caroline Bassett, Sarah Kember and Kate O'Riordan's *Furious: Technological Feminism and Digital Futures* (2020): in their last chapter, the one of the title, they make great propositions for going forward as voice(s) e.g., "To fall in with the claim that new is always better and that we need to replace the old, would be to forget that technology itself embeds, that older technologies continue to enact, even while attention has shifted to the new."

Michelle M. Wright
I'm a literature and literary theory professor, and personally, I have always found certain works of literature to be far more brilliant than anything a theorist—literary, cultural, political, social, anthropological, economic, scientific—has ever devised (and of course I think there is some brilliant theory out there)! I believe the best literature tends to be more nuanced, polyvocal, and dialogic—and daring—than theory. Right now, I am entranced, confused, fired up, and obsessing over Black women's contemporary speculative fiction, especially N.K. Jemisin's *The Broken*

Earth Trilogy and Nalo Hopkinson's novels *Midnight Robber* (2000) and *Sister Mine (2013)*.

I've been thinking more and more about how race is a heterosexual construct, how that impacts questions of Black women's agency, Black feminist politics, and, ultimately, belonging. If the primary duty of the race is its reproduction, where do Black women fit in outside the role of motherhood? I think many works in Black women's speculative fiction take up this question and find really astonishing answers, beginning with Black female protagonists who used to have clear roles within a patriarchy as wives and mothers, but now find themselves estranged, widowed, or otherwise violently launched into solitary status and a quest to find or maintain spaces where they are now agential. All the protagonists in these works are not only flawed but, in many cases, anti-heroes, and there seems to be a grimly abrogated pragmatics at work in these narratives, even as we are taken to very imaginative spaces and times.

The way that impacts race is equally intriguing, and reminds me of the Nella Larsen's rather brilliant treatise in her novella "Passing" (1929), which indicates that when racial binaries become ambiguous in the United States, so do economic, sexual and gender binaries—all of these flimsy ideas prop one another up in an effort to dictate what behaviors are normative and therefore acceptable, and what behaviors are dangerous and to be eliminated. We know race is not a biological category, but a historical one—people are grouped and labeled at some point in time for political and economic purposes. So what happens when a story takes place tens of thousands of years in the future or in the past, and your protagonist is Black? What makes them Black?

My last book, *Physics of Blackness* (2015), argues that Blackness is not a "what"—something fixed outside of time and space—but a "when" and a "where"—it achieves definition within historical contexts, however explicitly muted they may be. This current book project, *Afrotopias and the Agency of Space*, looks at how Black femmes from the 19th century up to the present day have theorized Black female agency, in travel narratives, essays, and fiction, and, in doing so, not only redefined the female, but race itself. So, in sum, I'm interested in what happens to Blackness when we read it outside of or in tension with the heteronormative.

Q7: Is There Anything Else You Would Like to Respond to, or Wish We Had Asked You About?

Laleh Mehran

Throughout my decades as a professor in higher education, I have been repeatedly asked to support DEIJ efforts, yet rarely has anyone asked if I need support, especially during devastating events related to Iran. My experiences have revealed a persistent conflation of equity and equality, which has eroded trust in administrators, institutional policies, and the university itself. This disconnect has also diminished my confidence that the energy and time I dedicate to my pedagogy are valued by the institution in any meaningful manner.

I'm increasingly hearing from colleagues and associates who have the means and are seriously considering leaving the country due to the current political climate. While their desire to leave is understandable, it raises concerns about intersectionality, as many marginalized populations lack the resources to relocate freely. This concerns me greatly, as it not only highlights systemic inequalities but also reduces the presence of those who could offer critical support (potentially in very serious situations) for DEIJ efforts.

I taught a course centered on the #WomanLifeFreedom movement, which gained global attention following the killing of Jina (Mahsa) Amini in 2022. Originally rooted in Kurdish women-led movements, the slogan Woman, Life, Freedom took on worldwide significance during the protests in Iran, sparked by Amini's death while in the custody of the morality police for alleged "improper" clothing violations. The feminist movement transcended gender, addressing intersectional identities, with issues of class, ethnicity, and social status overlapping. My students, many of whom are the same age as the Gen-Z protestors fighting for human rights, felt a strong sense of generational connection. They voiced frustration over the limited national news coverage and the lack of institutional support for Iranian students at our university. Inspired by the wave of powerful artwork emerging from Iran and beyond, they created their own artworks for a public exhibition. This solidarity continues, as students have since adopted similar resistance strategies to foster connections across national borders, recognizing shared challenges faced by their generation worldwide.

The heightened political pressure and violence, particularly against marginalized populations in the U.S., make it nearly impossible to focus or be productive. The constant fear of devastating consequences, such as state-sanctioned violence, criminalization, restrictions on bodily autonomy, or detention and deportation, promotes anxiety that stifles creative and intellectual output. This stress disproportionately impacts BIPOC women, as data shows a greater toll on our physical and mental health. The emotional and psychological strain of navigating a hostile political climate exacerbates existing health disparities, further hindering the ability to thrive or even simply survive in spaces meant to encourage growth and innovation. However, by embracing comprehensive intersectionality, we can start to alleviate these issues by building solidarity and advocating for structural changes that promote the holistic well-being and prosperity of all individuals.

Susanna Paasonen

Responding to these prompts has made me think about the presence, or role, of sexuality in cyberfeminism, which came around about the same time as queer theory (thinking of de Lauretis's 1991 text), and the degrees to which these ran parallel, and the degrees to which these intersected, and did not. The question of how the sex in cyberfeminist projects connected to sexuality more broadly would warrant proper research: I'm basically thinking/typing aloud here about the representational/textual bodies of cyberfeminist art, the contemporaneous enthusiasm about cybersex/teledildonics (which has definitely had a comeback with VR tech), and the spaces available for different kinds of sexual desires and orientations within this all. How deep did the queer routes meander? What did they discover? What has carried over?

While there definitely is an anti-sex strand to some current forms of queer theory, it's also been a realm for looking deeply into the political potentialities of corporeal joy as something that makes and unmakes bodies and connections. Which is very exciting.

About the Respondents

María Fernández, Ph.D., is an Associate Professor at Cornell University. Her research and teaching concern the history and theory of digital and new media art, Latin American art, and feminist media art, with attention to postcolonial/decolonial theories. She is the author of *Cosmopolitanism* *in Mexican Visual Culture* (Texas University Press, 2014), co-editor of *Domain Errors! Cyberfeminist Practices* (Autonomedia, 2002), and editor of *Latin American Modernisms and Technology* (Cornell Institute for Comparative Modernities and Africa World Press, 2018). She's currently writing a book on the work of the British cybernetician Gordon Pask.

Amelia Jones is Robert A. Day Professor at the Roski School of Art & Design at USC in Los Angeles, and is a curator and scholar of contemporary art, performance, and feminist and sexuality studies. Her 2021 book entitled *In Between Subjects: A Critical Genealogy of Queer Performance* explores the history of performance art and queer theory since the 1950s. She curated the recent exhibitions *Queer Communion: Ron Athey* (2021) and *Ken Gonzales-Day: History's "Nevermade"* (2025) and is currently working on a book addressing the neoliberalism and structural racism in the Euro-American university and art complex, *Lifework: Against Cultural Capitalism. Q*

Tania Kupczak, MFA, is currently the Creative Manager at Shout Your Abortion, an organization that elevates pathways to safe abortion access across the US regardless of legality. In her 50+ rotations around the sun, she has worked in the non-profit sector, academia, the theater and film industry, and is the owner of her own props and set fabrication business. As a visual artist and a witch, she really enjoys collaboration, plant medicine, and portals to the sublime.

Laleh Mehran, MFA, is a Professor of Art at Ohio State University. She creates elaborate environments in digital and physical spaces focused on complex intersections between politics, religion, and science. In a sociopolitical climate where certain views are increasingly sensitive and may have meaningful implications, Mehran's artworks are invitations to think again about each of these paradigms and the profound connections that bind them. Her research, often modeled on and about the very ideas of science and technology, takes advantage of their cultural importance in order to articulate a set of ideas that require precisely these kinds of mediations from such intolerances.

Susanna Paasonen, Ph.D., is a professor of Media Studies at the University of Turku with an interest in studies of networked media, affect, sexuality, and media theory. She is the PI of the research consortium, "Intimacy in Data-Driven Culture" (IDA, 2019–25), funded by The Strategic Research Council of Finland, and is the author of *Carnal Resonance: Affect and Online Pornography* (MIT Press, 2011), *Many Splendored Things: Thinking Sex and Play* (Goldsmiths Press, 2018), and *Who's Laughing Now? Feminist Tactics in Social Media* (with Jenny

Sundén, MIT Press, 2020), and *Yul Brynner: Exoticism, Cosmopolitanism and Screen Masculinity* (Edinburgh University Press, 2023).

Nell Tenhaaf is an electronic media artist living in Trent Hills, Ontario. Her artworks integrate elements from the biosciences and Artificial Life in the form of lightbox displays and interactive sculptures. She is known for her work in the 1980s that pioneered interactive graphics and text via experimental telecommunication networks, and her later works foregrounding feminist commentary on biotechnology and genetics. She is Professor Emerita at York University in Toronto.

Michelle M. Wright, Ph.D., is Augustus Baldwin Longstreet Professor of English at Emory University and the author of *Physics of Blackness: Beyond the Middle Passage Epistemology* (University of Minnesota Press, 2015), *Becoming Black: Creating Identity in the African Diaspora* (Duke University Press Books 2004), and co-editor of *Domain Errors! Cyberfeminist Practices* (Autonomedia, 2003).

References

Atanasoski, Neda, and Kalindi Vora. 2019. *Surrogate Humanity: Race, Robots, and the Politics of Technological Futures.* Duke University Press Books.

Baldwin, James. 1956. *Giovanni's Room.* Dial Press.

Bassett, Caroline, Sarah Kember, and Kate O'Riordan. 2020. *Furious: Technological Feminism and Digital Futures.* Pluto Press.

Beautiful Trouble. https://beautifultrouble.org/

Bernstein, Gaia. 2023. *Unwired: Gaining Control over Addictive Technologies.* Cambridge University.

Bilal, Wafaa. 2007. "Domestic Tension." https://wafaabilal.com/domestic-tension/

Black Trans Archive. 2020. "We Are Here Because of Those That Are Not." https://blacktransarchive.com/

Blas, Zach. 2011–2014. "Facial Weaponization Suite." https://zachblas.info/works/facial-weaponization-suite/

Center for Technological Pain. 2019. "DIY Solutions to health problems caused by Digital Technologies." http://centerfortechpain.com/

Chromicz, Zavisha. 2025. "Spilling Out." Paul Petro Contemporary Art.

Cinzia Arruzza, Tithi Bhattacharya, and Nancy Fraser. 2019. *Feminism for the 99%: A Manifesto.* Verso

Couldry, Nick, and Ulises, Mejías. 2019. *The Costs of Connection: How Data Is Colonizing Human Life and Appropriating It for Capitalism.* Stanford University Press.

Crawford, Kate. 2022. *Atlas of AI: Power, Politics, and the Planetary Costs of Artificial Intelligence.* Yale University Press.

Dihal, Kanta, Stephen Cave, and Sarah Dillon, eds. 2020. *AI Narratives: A History of Imaginative Thinking about Intelligent Machines.* Oxford University Press.

Dinkins, Stephanie. 2018. "Not The Only One (N'TOO.)" https://www.stephaniedinkins.com/ntoo.html

Faludi, Susan. 1991. *Backlash: The Undeclared War Against American Women.* Crown Publishing Group.

Ghani, Mariam. 2019. "What We Left Unfinished." Indexical Films Ajna Films. February 9. 1 hour., 12 mins. https://www.whatweleft.com/

Gomez, Jewelle. 1991. *The Gilda Stories.* City Lights Publisher. Haraway, Donna. 1985. A Cyborg Manifesto. Springer.

Hopkinson, Nalo. 2000. *Midnight Robber.* Warner Aspect.

Hopkinson, Nalo. 2013. *Sister Mine.* Blackstone Publishing.

Isaacson, Walter. 2014. *The Innovators: how a group of hackers, geniuses and geeks created the digital revolution.* Simon and Schuster.

Jemisin, N.K. 2015-2017. The Broken Earth Trilogy. 3 vols. Orbit.

Jones, Amelia. 2020. *In Between Subjects: A Critical Genealogy of Queer Performance.* Routledge.

Jones, Amelia. 2012. *Seeing Differently: A History and Theory of Identification of the Visual Arts.* Routledge.

Jones, Amelia. *Lifework: Against Cultural Capitalism.* Manuscript-in-Progress.

Keaney, Jaya. 2023. *Making Gaybies: Queer Reproduction and Multiracial Feeling.* Duke University Press.

Kenan, Randall. 1989. *A Visitation of Spirits.* Grove Press.

Klein, Naomi. 2023. *Doppelganger: A Trip into the Mirror World.* Farrar, Straus and Giroux.

Kupczak, Tania. 2002. "So You Wanna be a Cyborg Mommy? Queer Identity and New Reproductive Technologies." In *Domain Errors! Cyberfeminist Practices,* edited by María Fernández, Faith Wilding, and Michelle M. Wright. Autonomedia.

Larsen, Nella. 1929. *Passing.* Knopf.

Lauder, Adam. 2023. "Beyond Electronics." In *Variable Conditions: Paracomputational Arts in Canada.* McGill Queen's University Press.

Lauretis, Teresa De. 1991. *Queer Theory: Lesbian and Gay Sexualities.* Indiana University Press.

Lewis, Sophie. 2019. *Full Surrogacy Now: Feminism Against Family.* Verso.

Lorde, Audre. 1982. *Zami: A New Spelling of my Name.* Crossing Press.

Maggic, Mary. 2015. "Open Source Estrogen." https://maggic.ooo/Open-Source-Estrogen-1

María Fernández, Faith Wilding, and Michelle M. Wright, eds. 2002. *Domain Errors! Cyberfeminist Practices.* Autonomedia.

Najmabadi, Afsaneh. 2013. *Professing Selves: Transsexuality and Same-Sex Desire in Contemporary Iran*. Duke University Press.

Najmabadi, Afsaneh. 2005. *Women with Mustaches and Men without Beards: Gender and Sexual Anxieties of Iranian Modernity*. University of California Press.

Noble, Safiya. 2018. *Algorithms of Oppression*. NYU Press.

Paasonen, Susanna, and Jenny Sundén. 2020. *Who's Laughing Now? Feminist Tactics in Social Media*. MIT Press.

Perry, Sondra. 2016. "Graft and Ash for a Three Monitor Workstation." https://sondraperry.com/Graft-and-Ash-for-a-Three-Monitor-Workstation

Plant, Sadie. 1997. *Zeros and Ones: Digital Women and the New Technoculture*. Fourth Estate.

Ross, Loretta J., and Rickie Solinger. 2017. *Reproductive Justice: An Introduction*. University of California Press.

Russell, Legacy. 2020. *Glitch Feminism: A Manifesto*. Verso.

Sage-Femme Collective. 2008. *Natural Liberty: Rediscovering Self-Induced Abortion Methods*. Sage-Femme.

Salinas, Daisy. 2011. *Muchacha Fanzine*. Zine. Vol 1.

Seu, Mindy. 2023. *Cyberfeminism Index*. Inventory Press.

Shatz, Adam. 2024. *The Rebel's Clinic: The Revolutionary Lives of Frantz Fanon*. Bloomsbury Publishing.

She Who Sees The Unknown. 2016–2020. https://shewhoseestheunknown.com/

Shockley, Ann Allen. 1974. *Loving Her*. Bobbs-Merrill.

Shockley, Ann Allen. 2014 (1974). *Loving Her*. Open Road Media.

Smiley, Jane. 2010. *The Man Who Invented the Computer*. Doubleday.

Treister, Suzanne. 2009–2011. "HEXEN 2.0." https://www.suzannetreister.net/HEXEN2/HEXEN_2.html

Taylor, Brandon. 2020. *Real Life*. Riverhead Books.

Varoufakis, Yanis. 2023. *Technofeudalism: What Killed Capitalism*. Melville House.

Washington, Harriet A. 2008. *Medical Apartheid: The Dark History of Medical Experimentation on Black Americans from Colonial Times to the Present*. Vintage.

Wiener, Norbert. 1948. *Cybernetics: Or Control and Communication in the Animal and the Machine*. MIT Press

Wilding, Faith, 2002. "Rant of the Menopausal Cyborg." In *Domain Errors! Cyberfeminist Practices*, edited by María Fernández, Faith Wilding, and Michelle M. Wright. Autonomedia.

Wright, Michelle M. 2015. *Physics of Blackness: Beyond the Middle Passage Epistemology*. University of Minnesota Press.

Wright, Michelle M. *Afrotopias and the Agency of Space*. Manuscript-in-Progress.

Yusoff, Kathryn. 2018. *A Billion Black Anthropocenes or None*. University of Minnesota Press.

CHAPTER 4

Magical Life

Nathanael Elias Mengist and Irina Aristarkhova

In *Magical Life*, we ask "Whose/what life is magical?" As a way of answering this question, we offer our visions of art and life practices inspired by: a) subRosa's worldbuilding work of "undoing gender" in *Yes Species*, after which we present Mengist and Ioan Buitu's *DIY Homunculi* and plant Assisted Reproductive Technologies (*pARTs*); b) history of alchemy and its manifestations in postcolonial technoscience studies (Rajani Sudan) and Black Feminist Thought (Sylvia Wynter, Dorothy Roberts, Zakiyyah Iman Jackson, Tiffany Lethabo King); c) reproductive mutual aid of Black kinship (Cynthia R. Greenlee, Dorothy Roberts, Jennifer Morgan, and Alys Eve Weinbaum) and Argentinian "abortion joy" (Lea Happ); and d) subRosa's *Sex and Gender in the Biotech Century* performance with its options for the audience to choose their future genetically modified offspring from decision trees, which will culminate in our own visions of *Indecision Trees* of probability logics. Throughout the chapter, we center the collective worldbuilding of life within the effort of *compassionate inquiry*.

A Compassionate Inquiry

Reproduction and creation of new life are the primary sites of operations of power struggles and power structures. Control over life and death assures those in power of their own reproduction and their power's maintenance. In addition to addressing existing biotechnologies, this chapter builds on

H. Willis et al., *Night Sweats: Cyberfeminist Practices*, Palgrave BioArt,
https://doi.org/10.1007/978-3-032-06955-9_4

subRosa's works and texts that imagined what a new world might look like. subRosa drew on the history of witchcraft, alchemy, and women's resistance through communal sharing and traditional knowledge(s), passed from one generation to the next, through doulas, midwives, and women's religious and spiritual societies. They were also inspired by feminist and AIDS health movements of the 1970s and 1980s, when communities organized to serve themselves because the biomedical professions either abandoned or exploited them.

In this Chapter, we extend these forms of alternative worldbuilding to where we can find them today, both in imagination and in community organizing around "reproductive mutual aid." We see the flourishing life of those who have primarily been used as instruments towards others' goals as affirming and empowering all life. We consider embodied practice within each other's professional and wider communities to also be a work of imagination towards birthing and sustaining new worlds. In this work, each location presents its own challenging histories that need to be addressed, due to the way in which the above-mentioned structures of power have been amassing and entrenching exploitative and extractive operations. Even where the desire for life is harnessed towards extraction and privatization, we find magical life being possible and happening all around us. When working collaboratively on this chapter, we challenged ourselves to remain open to what we called a *compassionate inquiry*. We wanted to acknowledge the real needs and desires of citizens and communities that scientists, doctors, investors, bioethicists, lawyers, and industry workers evoke and respond to when they develop and implement Assisted Reproductive Technologies (ARTs), and are involved in cutting-edge research. We wanted to acknowledge the heartache and inequalities built into ARTs and reproductive (in)justice in general, as we bring together seemingly contradictory layers of opposing reproductive goals. We want to acknowledge deep desires and longing, often in danger of being not only exploited by profit- and control-driven extraction and privatization of birthing and maintaining a life, but also of being dismissed by some scholars and activists because of ARTs' connection to those capital flows and exploitative practices. And we wanted to focus here on specific pathways to strengthen "the weakened" (see Chap. 1) through self-empowerment and creative worldbuilding. We do not claim that we achieved this compassionate inquiry; this is not for us to judge, but we wanted to share it as our guiding goal when writing this chapter.

Our vision of expanding the angle of feminist and cyberfeminist focus to the realm of complexities and contradictions of compassionate inquiry also responds to María Fernández's call in this volume (see Chap. 3). Fernández cites feminist technoscience scholars of the recent past who show how wider perspectives that are raised internationally about ARTs should also include children who are born as their result. The first generation is grown up now and sometimes starts questioning their parents' choices and the biomedical systems that brought them to life:

> Now there are more options to create families than ever before from ARTs to family selection based on affinities. After Keaney, I suggest that we consider the consequences of our choices in broader social and political spheres. What effects do they have on our immediate families and in our communities? Who or what do they impact further away and how? Atanasoski and Vora's historical examination of surrogacy in relation to robotic technologies and race reminds us that technologies are never free from culture and invites us to think about people in various contexts who support reproduction in the wider sense. (Fernández, 77, this volume)

Our interest in this expanded view builds on our previous work (see, for example, Aristarkhova 2012, on hospitality of the maternal, ectogenesis, and trans- and male- pregnancies, and on contradictions of feminist judgment of ARTs, see Aristarkhova 2018; and Mengist 2023). Specifically, this chapter brings forth our joint interest in these topics that had started from conversations and panels at an annual conference of the Society for Literature, Science, and the Arts (SLSA), at the University of California, Irvine, in 2019. In terms of the distribution of our writing effort, Mengist's contribution to the Chapter is primarily centered on sections related to *Alchemy* and *Reproductive Mutual Aid*, while Aristarkhova's is on subRosa's works, *Yes Species* and *Sex and Gender in the Biotech Century*, and, to a lesser extent, *Reproductive Mutual Aid*. The rest has been written collaboratively.

Yes Species and Creative Worldbuilding

> Imagine ... a clearing in the forest is populated with a living montage of becomings and unbecomings. Greetings and other exchanges transpire in a semi-digital staging of interruptions.

> Imagine … in a clearing in the forest, a symposium of erotic philosophers exorcises gender sickness. Witches, athletes, herbalists, and yes species moan, choking on the silver apples of the moon, the golden apples of the sun. (subRosa 2008, 235)

These opening words are chants to create a new world for beings who do not conform to a biomedical binary sex and gender system, especially those who had surgeries or other interventions to "correct" them when they were born—intersex people ("Hermaphrodites"). They are part of subRosa's collaboration with Wu Tsang, a cross-disciplinary artist who makes narrative and documentary film, live performance, and video installations. In 2005, subRosa was invited to present an opening performance and installation for an exhibition and archive project called "1-0-1 [one 'o' one] intersex: The Two-Gendered System as a Human Rights Violation" at the New Society for Visual Arts in Berlin (NGBK, Neue Gesellschaft für Bildende Kunst), and run by Ulrike Klöppel, Wijnand Koppelmann, Ins A Kromminga, Nanna Lüth, Rett Rossi, and Karen Scheper de Aguirre. subRosa invited James Pei-Mun Tsang (who used the pronoun "he" in 2005). Today, the artist is known as Wu Tsang and uses the pronoun "she," which we will use throughout this chapter. Together, they produced *Yes Species,* a "performative tableau and installation" that has been presented as a kind of "uncommon coupling" that will be affirmative in offering "every possible explanation of yes species" (subRosa 2008, 235). A book (subRosa and Tsang 2005) was also a part of the project and is discussed further in this Chapter.

Yes Species engaged with the exhibition's mandate to "serve a political purpose to convey knowledge about 'other' worlds and to defy expectations and definitions about what art about intersex might be" (ibid). Here, subRosa and Tsang's performance used the form of a "poetic and affective tableau" to "evoke feelings and ideas by appealing to the senses and imagination through colors, sounds, images, and texts" (ibid.). In the archival documentation video of *Yes Species* performance in Spain, we watch as Wilding and Willis slowly move on stage, covered by a multicolors forest and text on the wall behind them and on long vertical stripes of paper arising from the ground (unpublished video documentation, Hyla Willis personal archive, watched June 2025, see Fig. 4.1 and Fig. 4.2). This video projection of a vividly green new-leafing forest creates a sense of a "forest clearing."

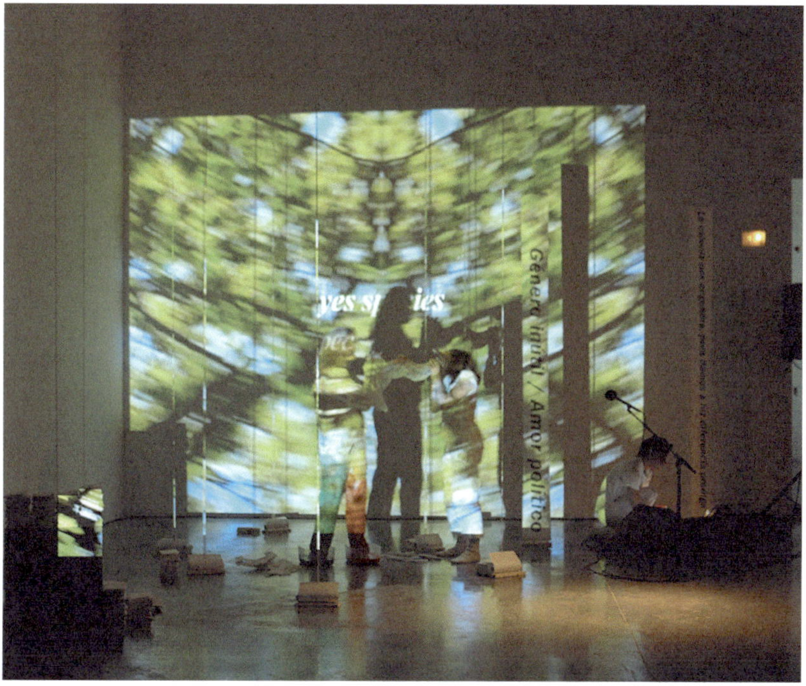

Fig. 4.1 *Yes Species* performance video still depicts, L to R, Hyla Willis, Faith Wilding, and Wu Tsang. Willis and Wilding are jointly inflating a vellum "word balloon" while Wilding's feet soak in vats of dye, and Tsang performs a live DJ mix of ethical vocalizations (subRosa and Tsang 2006, Spain).

For subRosa, multiple formal choices were meant to problematize gender appearance expectations and stereotypes. Tsang, playing the role of an on-stage DJ, positioned herself in the middle of the "forest" to play:

> mixed songs, vocalizations, and sounds from trans, queer, and female vocalists. Brechtian interruptions occurred when a second performer operated bilingual scrolling texts that rose up from the floor, declaring, for example: "The human body was the first machine developed by capitalism," "Things could be thought differently," "I love to you," "Andere Handlungsweisen," and "so beautiful, so various, so new." A third performer, costumed like mad King Ludwig of Bavaria, stood with his/her feet in containers of red and green ink, breathing and vocalizing into human-organ-shaped vellum

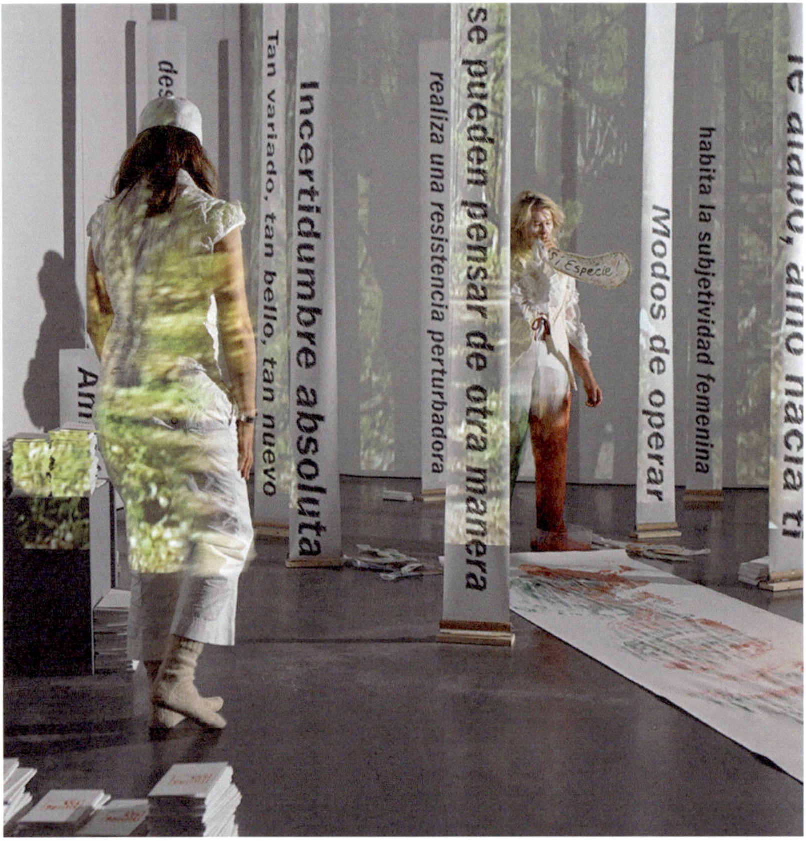

Fig. 4.2 *Yes Species* performance video still depicts Willis (L) unfurling vertical banners and distributing the *Yes Species* booklet while Wilding (R) is inflating a vellum balloon and "painting" a floor scroll with her dyed feet. Tsang (not pictured here) is performing a live sound mix on stage. (subRosa and Tsang 2006, Spain).

> balloons, and later painted a long scroll with red- and green-dyed feet and dripping pants legs. Meanwhile, the second performer hand-stamped the cover art for a book of texts written to accompany the performance, and then distributed the books among the audience. (subRosa 2008, 232–233)

The work was performed the following year, in 2006, as part of the exhibition and festival "Cyberfem: Feminisms on the Electronic Landscape," at Espai d'Art Contemporani de Castello, Spain. The book accompanying these performances, *Yes Species,* (subRosa and Tsang 2005), contains salient, historically important texts that developed diverse approaches to "undoing gender" and other binaries, including those they considered to be blind spots in queer communities. This work opens a discussion about reproductive justice for those who were until recently excluded from notions of parenting and were denied access to ARTs, showing how, for subRosa, questions of bioart and their practice included these communities as a question of "political love." In *Yes Species*, subRosa and Tsang focused on reproductive rights and futures for LGBTQIA+ persons, opening a wider conversation about justice:

> Since the 1990s, another strong challenge to the notoriously unjust, racist, and sexist U.S. health care system has come from gender-queer, trans (transsexual), and intersex people who must contend with biomedical practices, human rights policies, and legal institutions in many different ways. Genderqueer and trans activists are contesting a wide array of medical, cultural, and disciplinary systems. Borrowing tactics from the FHM, ACT-UP, and queer activism, intersex/trans activism addresses questions of gender difference, race, and sexual, reproductive, and civil rights that are at the heart of many cultural, political, and human rights struggles. Consequently, trans and queer activists' campaigns for the right to bodily autonomy and choice of gender identification and biological sex could be as significant in bringing about profound legal, political, and societal changes in the twenty-first century as the civil rights, feminist, and gay/lesbian movements of past decades, to which they are intrinsically linked. (subRosa 2008, 232–233)

Here, almost twenty years ago, SubRosa and Tsang make bold claims for the new world and call for the annihilation of gender as we know it. We hear the sounds of this new world, of a magical becoming that needs to happen through self-rebirth. These visions are different but also connected to the visions on alchemy of the *Sex and Gender in the Biotech Century* project illustrations, which are discussed below.

What does this new world specifically look like, and why do we need to undo the gender binary? In her text in the book, "Vocalization in an Ethical Matrix," Tsang suggests "becoming-women" as "a declaration of new feminist work" in order to "get somewhere, to progress (an unfashionable ideal), to get specific about commonality, and not to hide in the realm of impossible negative" (Tsang 2005, 16). This gesture, among others that Tsang proposed, centers her worry that "the trans community is becoming cast as boy-centered. I am confronted in my community by the authoritative presence of whiteness, privilege, and unproductive victimization" (Tsang 2005, 13). Here, whiteness, privilege, and unproductive victimization are connected to something "authoritative," almost becoming authoritarian in their contextual meaning, and calling for a tactic to counteract the authority. Tsang's performance and this co-authored text with subRosa help us understand how her careful approach "with respectful / regards for authority" (listed as legitimate categories, norms, ruling figures, athletic prowess, mainstream media, etc.) does not shy away from facing the terms of violence, "which can deny/obstruct ...viable life at the levels of the body, citizen, and state," listing—"to perform connectivities,"—"rape, invasion, coercion, hate, violation, imperialism" (Tsang 2005, 12).

Thus, violence becomes a conjoined twin of authority, just like whiteness becomes an institution, protected by the army, the police, the nuclear family, and other institutions of power, of "strengthening" (see Chap. 1). In this project, Tsang and subRosa's cyberfeminist moves and innovations form a new type of partnership that is undoing the heteronormative and human-centered from below, growing from the ground in this new forest, and also refusing the neat separations between organic and non-organic, or human and non-human. Who possesses the magic of life here? Those who can imagine new magical lives that are outside of the coercive systems with their imposed significations, force, and violence.

subRosa and Wu Tsang practice what they call "uncommon couplings," modelling it for us. In this new world, the powerful and the disempowered (or "strengthened" and "weakened," as discussed in Chap. 1), are also not just opposed. We should not expect minoritarian life to be eclipsed by others' hatred and violence, as institutionalized, structural, or omnipotent as it seeks to be in order to instrumentalize and discard. Thus, as Tsang describes the importance of the performative and the specific, so as not to lock ourselves in any kind of "absolute statements" that promise their own authority, she argues that it would also be a mistake to maintain "a general expectation that minority artists [groups] should formulate their work in opposition to the forces that oppress them, rather than in any other way imaginable" (Tsang 2005, 20).

Here, Tsang deconstructs de Certeau's oppositionality between the strong and the weak, showing how not everything is a tactic, that there are spaces where one creates, imagines, and plays, and this could be an ongoing strategy of magical life, undecipherable, no matter how much it is studied and manipulated, in vivo and in vitro. Tsang then, in her alchemic gesture of metamorphosis, bestows authority to the person who enunciates, on what is "avant-garde" and what is not, or who is "white," or not. Tsang declares: "To recapture the stagnancy of this backwards formulation, I say avant-garde belongs to me, and I am not white—pornography belongs to me, and I am queer" (Tsang 2005, 20). Later in the text, evoking the influence of Donna Haraway's "Promises of Monsters" on her thinking, Tsang remembers that it was written the same year when Tsang listened to *Bikini Kill* "(who ten years later is now the reincarnate *LeTigre*)" (Tsang 2005, 24), bringing this specific, personal, history, into the stance of defiance in relation to subaltern form that non-Western, non-white, queer, artist is supposed to enact: "Here, I collapse boundaries between linguistic debates and queer politics, hoping to synthetize strategies that help us to bring it all together, without reducing every particular. The shape grows rounded and dense. ... The word 'ethnic' becomes leveraged, in part as an old reclamation trick, to take back what multiculturalism glossed over as being insensitive" (Tsang 2005, 21–22, 23).

This complex, collaborative performance, or performative collaboration, takes place within impossible alliances and music scores (see Fig. 4.3) that come from "all over the place" for each performer. But they are also from everywhere and nowhere—their own unique voices, mixed across machinic and collaborative intentionalities and accidental choices.

When Tsang asks: "Does to 'break free' translate into full autonomy?" (Tsang 2005, 21), subRosa answers with their own text in the *Yes Species* book, "*Useless Gender: An Immodest Proposal for Radical Justice*," where biotechnology, witchcraft knowledge and midwifery are reclaimed for the contemporary productive/reproductive self-determination. With affinity to Tsang, subRosa wonders what would happen:

> if we would refuse to perform gender in any way any longer. What if we made gender useless to the "labor" of the reproduction of femininity and masculinity and the nuclear family? ... In other words it would no longer be compulsory to be female or male in any socially constructed (or biological) terms – you could be "yes species" and make up what that means as you went along, imagine. ... Thirsting for justice and joy in the practice of useless gender, we freely choose for a political love, a social love transcending privatized gendered love. (subRosa 2005, 57–58)

Yes Species was born in 2005, "thirsting for justice and joy," that are central, as a couple, to our conception of Magical Life here. Since that moment, two decades have passed, and the fruits of that queer and trans

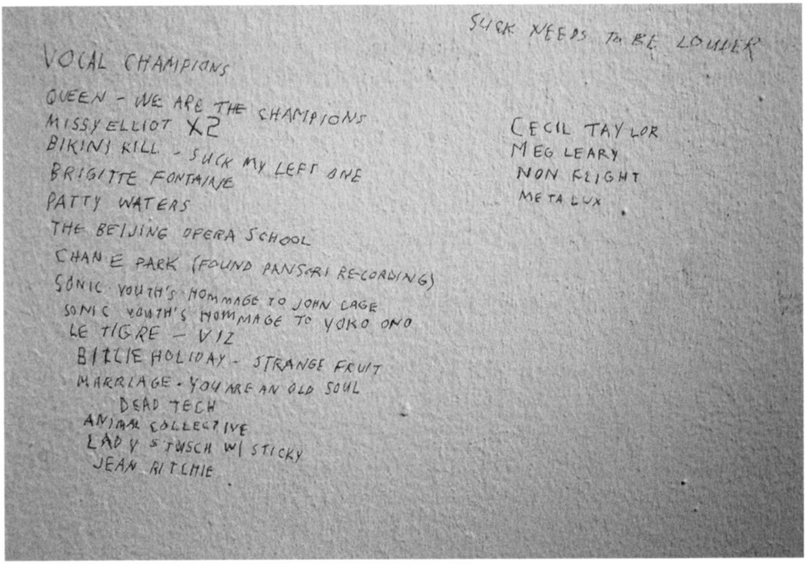

Fig. 4.3 *Yes Species* detail. A set list for the live sound mix by Tsang, handwritten on the wall (subRosa and Tsang 2005, Germany).

activism have brought about “profound legal and societal changes” just as subRosa and Tsang predicted and enacted in *Yes Species.* The current backlash against LGBTQIA+ persons is a testament to these changes that also seeks to undo them (see Chap. 3). Tsang and subRosa’s experimentation and collaboration (with performances in Germany and Spain, and the publication of their writing and texts) is a legacy that has continued to raise questions that are still with us today: whether sex/gender split serves cyberfeminist and feminist interests any longer, whether women/men terminology makes sense anymore, to whom and why, what kind of labor they perform for society, economy, politics, culture, for whom, and who and how much are invested in them. Tsang’s work has also raised questions about “ethnic” differences and policing of gender and sexuality cross-culturally and globally, highlighting the challenges that artists face in relation to other identity markers that have a direct impact on viability of their lives and creative practice: resources, mobility, economic and personal security.

subRosa's wider work has shown that the semantic, grammatical, and biological imperatives of sex and gender do a lot of heavy lifting for sense-making and meaning-producing in human societies that rely on them. Could magic and wisdom fill the void left by institutionalized extractive knowledge in our souls, drained of all life by efficient reproduction of binary worlds, including in biology and biomedical engineering that inscribes digital binary into biological "systems"?

Hence, bringing back the issues of ancient knowledges, magic, and witchcraft, becomes part of cyberfeminist tactical biopolitics:

> . . .Witches becoming-animal flying high on herbal magic
>
> . . .Monstrous cyborg graftings of plant, animal and machine refusing the essentialism of a gendered body
>
> . . .Daphne turning into a laurel tree to escape the embraces of Apollo
>
> . . .The female mystics in the Middle Ages marrying a virtual Christ and dying of spiritual ecstasy, rather than have their bodies controlled by male priests or husbands
>
> . . .The priests of Cybele sacrificing their testicles to the service of the goddess of life
>
> yesyes
>
> . . .Men performing femininity (not female impersonators), playing women's parts in Chinese opera or Shakespeare's plays
>
> . . .Sacred harlots (all sexes) pleasuring all strangers in the temple grounds
>
> . . .Women/men/queers choosing not to bear or rear their own children but to cultivate mentorship of other people's children
>
> . . .Queer affinity groups sharing resources, knowledge, sexual pleasure, and sometimes children." (subRosa 2005, 57–58)

Here, we are particularly interested in this aspect of informed, new, revisionist, affirming worldbuilding that *Yes Species* performed (even if we do not agree with all of the aspects of their new world). In fact, before encountering subRosa's work, one author, Mengist, created an artwork titled *DIY Homunculi* that was made with his former student and friend Ioan Butiu, that also evokes plant assisted reproduction. It debuted at the 2019 SLSA Conference and has since grown into a multimedia worldbuilding project that is included in the print of a grassroots speculative fiction zine and an arts anthology *The Anthropocene Laboratory* alongside an essay on the topic (Mengist 2023). This worldbuilding is collective and the DIY aspect is part of the intention for a grassroots movement (see Fig. 4.4).

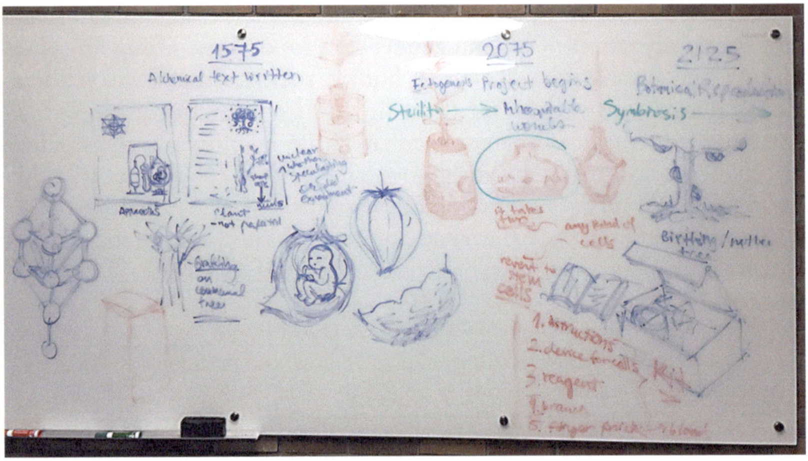

Fig. 4.4 A whiteboard drawing detailing the prototyping of DIY Homunculi (Ioan Butiu and Nathanael Elias Mengist, 2019).

DIY Homunculi is featured in an animated video podcast called *Life Touching Life* (2023). In the imagined world of *DIY Homunculi*, humanity has seen declining birth rates everywhere outside of Sub-Saharan Africa, so surviving communities must come together to develop a means of *plant-assisted reproductive technology* (**pART**). The story imagines reproductive freedom for all queer and trans people to make babies with each other through a partnership with plants. This is accomplished through the invention and application of the "matrixial vase" (See Fig. 4.5), borrowing the term from Aristarkhova's book *Hospitality of the Matrix: Philosophy, Biomedicine, and Culture* (2012). The device, depicted in an illustration, 3D print, and video, represents an Afrofuturist reclaiming of conventional ARTs that mirrors the mutual aid approach explored later in this chapter. Here is an excerpt from *DIY Homunculi* that describes the pART process:

> By exposing peripheral blood samples from two or more parents of any sex to a specially formulated reprogramming factor, certain hematopoietic stem cells can become induced with pluripotent stem cells, ultimately developing into viable gametes. These hybrid haploid cells are then absorbed by an Eden willow cutting within the matrixial vase, where they

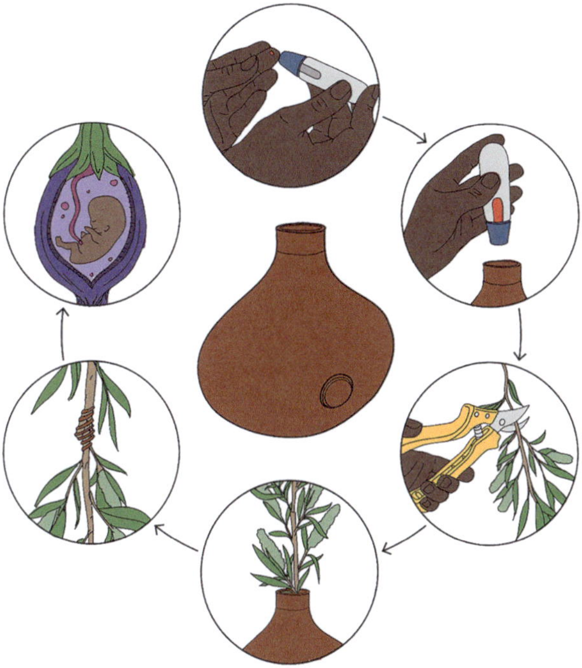

Fig. 4.5 Nathanael Elias Mengist and Ioan Butiu. The Matrixial Vase at the center of the process for Plant-Assisted Reproductive Technology (*pART*) from *DIY Homunculi*. Illustration by Ioan Butiu. 2019.

> are fertilized and incubated from the convenience of the home. Once this 'impregnated' cutting is grafted back onto the Eden willow from which it came, the tree will require intensive care. If all goes well, the lucky parents may welcome their child into the world in roughly nine months. (Mengist 2023, 160)

It happens that in a magical moment of life and creative intersections between subRosa and Tsang's own inspiration from Donna Haraway's work, Donna Haraway was at the SLSA Conference in 2019, on the occasion of her receiving SLSA's Lifetime Achievement Award. During her interactive speech, when she asked audience members to share their "carrier bags," Mengist came on stage to publicly present the matrixial vase to the audience (See Fig. 4.6). Fittingly, in her acceptance speech, Haraway called for work that centers both environmental justice *and* reproductive justice, which *DIY Homunculi* had done.

One image that was made as concept art for *DIY Homunculi* and was not included in any of the final versions of the project has a striking similarity to an image that subRosa created for one of their works, *Sex and Gender Education Show* pamphlets, which shows a pregnant, presumably white woman within a flask surrounded by iconography (See Fig. 4.7).

Co-creators Butiu and Mengist also conceptualized an alchemical emblem of a Black pregnant woman who is depicted on a scroll as part of a mystic configuration of lines and symbols (See Fig. 4.8). In the lore of *DIY Homunculi*, she is part of the prophecy that led to the invention of the *pART*, inspired by the apocryphal Jewish figure of Lilith and of Christian depictions of the Black Madonna. Mengist and Butiu were also inspired by some of the most well-known alchemical emblems from the 17th century, *Atalanta Fugiens*, which feature animal, celestial, and hermaphroditic icons as part of its transmission of encoded instructions for laboratory procedures. These icons would be at home in *Yes Species* and the exhibition devoted to intersex lives.

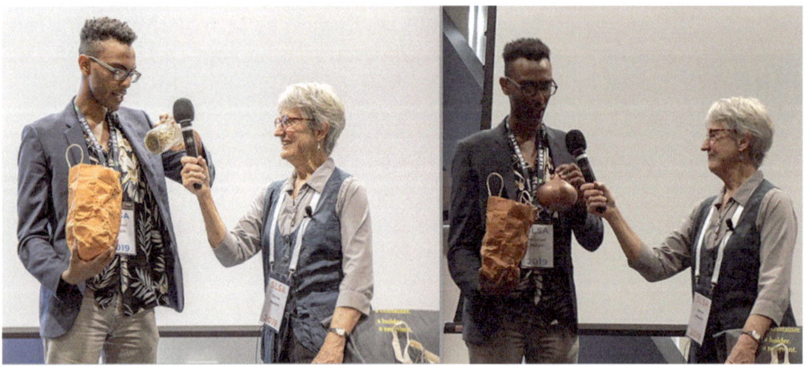

Fig. 4.6 Two images side by side. The left image depicts Nathanael Mengist with Donna Harraway about to reveal the Matrixial Vase during the SLSA Conference, Irvine, 2019. Courtesy of Jesse Jackson, 2019. The right image depicts Nathanael Mengist revealing the Matrixial Vase to Donna Haraway at the SLSA Conference in Irvine, 2019. Materials: 3D print, copper chrome spray paint. Photograph by Jesse Jackson, 2019. Courtesy of Nathanael Elias Mengist.

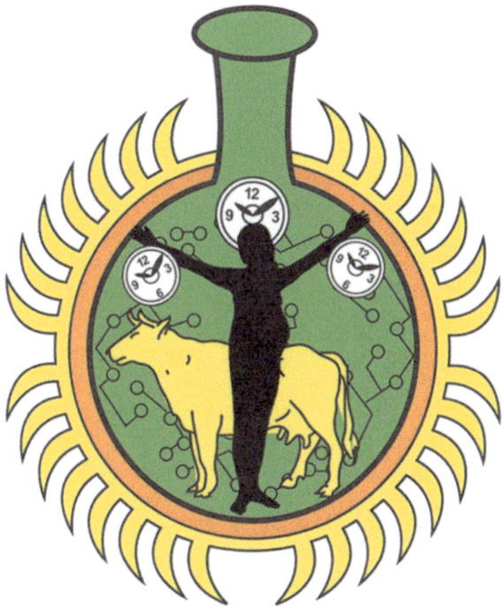

Fig. 4.7 An alchemical emblem collaboratively created by subRosa for the *Biopower Unlimited!* Map, website, and costumes (subRosa 2002).

Alchemic Reproductive Histories and Speculations

Alchemy allows us to explore issues of power in unexpected ways. Abandoned only in name, the dreams of the alchemists are alive and well. By recovering their stories, we further develop cyberfeminist insights about assisted reproductive technologies while bringing them into conversation with contemporary debates and desires.

Alchemy can be thought of as an early scientific knowledge system that integrated design traditions from across North Africa, Asia, and Europe (Sudan 2004). Further, the alchemical reproduction of life-forms in medieval and early modern imaginaries contains numerous antecedents to our contemporary debates about artificial intelligence, bioengineering, and ARTs (Newman 2005). One sixteenth-century European alchemist, Anna Maria Zieglerin, even offered a subversive vision of the creation of a homunculus by attempting to augment her own reproductive body (*in vivo*) rather than relying on *in vitro* techniques (Nummedal 2001), as expanded on below.

The early feminist research on these topics was focused on historiographic windows into the misogynist impulse to replace women's wombs with artificial ones. In *The Mother Machine: Reproductive Technologies from Artificial Insemination to Artificial Wombs* (1985), Gena Corea

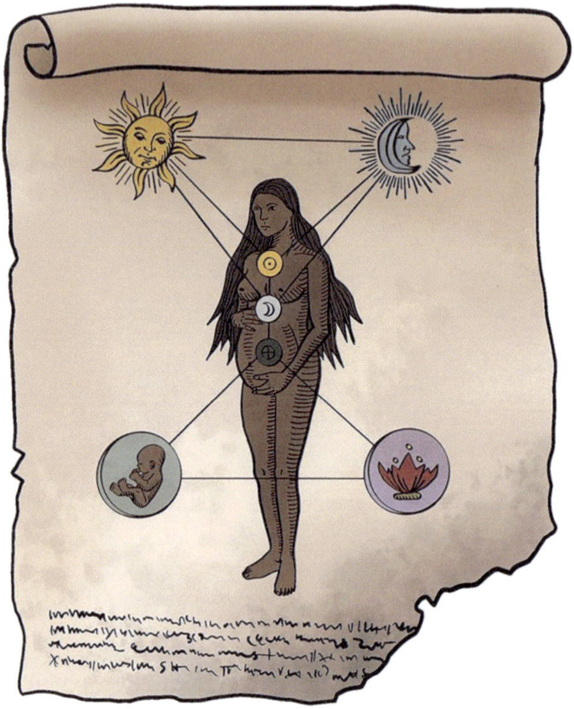

Fig. 4.8 Sketch for *DIY Homunculi*. Concept art by Ioan Butiu, 2019.

made a direct link between alchemical reproduction and the fertility industry:

> The notion that a man might create a human being by himself, without a woman, is an ancient one embodied in both the legend of the golem of medieval Jewish mysticism (a clay creature that had been given life), and of the homunculus or diminutive man, alchemists attempted to create. Boys' urine, blood, and sperm were among the ingredients alchemists prescribed for the creation of a little man.... This is the classic patriarchal myth of single parenthood by the male. Cloning promises (perhaps falsely) to bring the myth to life. (Corea 1985, 260)

As disturbing as the ingredients for making a homunculus may sound today, there are even stranger instructions for the alchemical creation of basilisks—mythical chimeras with the body parts of both a rooster and a toad. According to William R. Newman, while both Arabic and Latin alchemists dreamt of the masculine homunculus as an achievement to be applauded, by the 16th century the specter of the basilisk came to represent the "dreaded" feminine counterpart: "its masculine twin, the homunculus, is 'a miracle and magnale [mighty work] of God,' the basilisk is a 'monster above all other monsters'" (Newman 2020, 45). In at least one text, the key ingredient for making an alchemical basilisk was human menstrual flow (Newman 2020, 37–38). It is important to note that this is the ideological background from which another alchemist would subvert these problematic visions of what Tara E. Nummedal calls "alchemical reproduction" (2001). Nummedal, whose research focuses on the cultural history of alchemists and artisans in the Holy Roman Empire, describes a unique reproductive imaginary expressed through the biography of the German alchemist Anna Maria Zieglerin (c. 1550–1575), who claimed that through alchemy she could bear a child every 4 weeks, and that "the children would live until the end of the world and never get sick" (Nummedal 2001, 62).

Zieglerin is a valuable case study for cyberfeminist thought and practice today. Zieglerin helped run a laboratory where she experimented with a process that she claimed could generate new forms of life through her own womb. She intervened in contemporaneous discourse about the alchemical creation of homunculi for the purposes of knowledge and instead proposed that her art be an act of moral duty in the face of the coming apocalypse. She pursued questions about the possibilities of harnessing the power of what we might think of as artificial life and ARTs today, where a self-presenting woman technologist is hacking stereotypes and augmenting her own body to become a surrogate for the salvation of humanity.

Zieglerin's story also sets the stage for today's debates about the beginnings of human life and Christian ethical interpretations of ARTs, which we will return to later. It also presents a convergence with some of subRosa's criticisms about the growth of the ART industry and broader feminist technoscience studies arguments against "global fertility chains" (Vertommen, Parry, and Nahman 2022)—namely the drive to generate an ever increasing market of excess human reproductive matter (see also subRosa's *U-Gen-A-Chix* and *International Markets of Flesh* and *Expo EmmaGenics*, discussed further below and in Chap. 6). In the early days of European capitalism, Zieglerin's lab work was dependent on patronage and deliverables—so much so that her life was on the line. We imagine Zieglerin as an early bioartist, welcomed into the space of *Yes Species* with her experiments. Finally, Zieglerin's story, though she was "hardly a feminist" according to Nummedal, is still useful for imagining what cyberfeminist biopolitical practice could have looked like in the past (2021, 131).

There is more to alchemy, in addition to its aesthetic value. Corea, Newman, and Nummedal show that there are substantive reasons for considering alchemical theories of reproduction. All three researchers highlight the conventional views of reproduction that many alchemists intensified—masculine essence seen as superior to feminine substance. Within this masculinist vision of alchemical reproduction, Zieglerin believed in another vision for how alchemical practices could subvert the limits of what her reproductive body was capable of. This nuance is belied by visual gestures toward alchemy or simplifications of alchemists as figures. Alchemical reproduction today represents more than just myths coming to life—the dreams of augmented birth that alchemists participated in are alive not only in the capitalist desires to control and monetize ARTs but also in the radical dreaming of Black women and Black birthing people who champion more equitable access to ARTs (e.g., an organization called "Fertility for Colored Girls + Guys," 2025, that will be discussed later).

Alchemy and Transmutation: Who Is to Be Created

As in Zieglerin's case, subversive visions of ARTs are emerging today. Here, a critical historicization of ARTs, which Dorothy Roberts (2011) calls the "High-Tech fertility industry" and Jennifer Morgan and Alys Weinbaum (2024) call "reproductive racial capitalism," goes hand in hand with our desire to develop a compassionate inquiry and new terms of debate. Specifically, we are interested in the pragmatics of Black birthing people's everyday decision-making about ARTs in light of reporting by Cynthia Greenlee (2023). But first, before we turn our attention to new reproductive health movements through sharing tissue and biological materials from the infertility industry, we want to bring in a new approach to alchemy as a bridge to contemporary examples. While focusing on some aspects of the topic here, in a presentation at the Society for Social Studies of Science (4S) conference, Mengist (2025) brought together ARTs, alchemy, and the birthing of Black babies by employing a modification of André Brock's (2016) "critical technocultural discourse analysis" to examine the beliefs and practices encoded in "Enclosures: Blackness and Transmuation," a Black feminist online workshop hosted by the *Practicing Refusal Collective* (2021) and set in a broader context of prominent Black feminist scholars referencing alchemy in peripheral but powerful ways.

To start, Rajani Sudan provides a helpful model for how to reveal alchemy's role in relations of power in her 2016 *The Alchemy of Empire*. Specifically of use here is her description of alchemy as "a paradigm of reason that turns base material into something with infinitely more value" (Sudan 2016, 6). In Sudan's research, the epistemic resources of alchemy assisted the British East India Company in its appropriation of Indian material sciences. Opaque technological processes, such as the production of plaster of such high quality that it seems more like marble, were "sublimated through alchemical language into transparent, enlightened, scientific truths" that can be put into service for the building of the British Empire (Sudan 2016, 6). Sublimation—an operation used by both the earliest alchemists and the chemists of today — stands in for transmutation, which is more like that process of turning base matter into infinite value.

In Black Studies, there has been a similar approach to revealing power relations. A key example is Silvia Wynter's (2003) conceptualization of the overlapping phases of Western modes of self-representation across millennia—from the star-gazing Pagan and theistic Christian to the "liberal Man1" and the "biocentric Man2." In Wynter's vision, the West's self-conception as the "apex" of humanity is enabled by subordinating Black people to the absolute lowest state of inhumanity, where the Negro is the "nadir" (Wynter 2003, 300). Though never made explicit in Wynter's essay, both Zakiyyah Iman Jackson (2021) and Tiffany Lethabo King (2019) use the term "transmutation" to describe Wynter's "sociogenetic" theory of Western humanism as cumulative stages in a process of Human alchemy. Similarly to Sudan's use of "sublimation," King and Jackson also add their own chemical metaphors to their interpretations of transmutation—"residue" and "catalysis," respectively—showing how generative alchemical discourse can be for epistemic exploration. However, they use transmutation to reveal opposing processes.

For King, this discourse reveals how Wynter is mapping the "shifts and epistemic ruptures and the areas in which there is overlap between, or residue (a transmutation) that carries over" from one humanism to another (King 2019, 16). For Jackson, who is commenting on Wynter's aesthetic practice of "decipherment," transmutation points toward how this alternative to critique "catalyzes a transmutation [read: overturning] of Man as a speaking/knowing/feeling being" (Jackson 2021, 73). Here we can observe at least two distinct uses of transmutation in interpreting Wynter's ideas: (1) a negative transmutation that contributes to the ascendance of Western domination, and (2) a positive transmutation that contributes to the overturning of Western domination.

Most interesting for this chapter is how Black artists and educators find transmutation to be generative of affirming experimentation with Blackness. An exemplar is in the description of "Enclosures: Blackness & Transmutation," where participants "draw upon concepts of futurity and imagination as modalities that transmutate enclosures into other times/spaces/things, into liberation" (Practicing Refusal Collective 2021, n.p.). This is the second, positive form of transmutation in Black Studies, which counters anti-Blackness with what Kevin Quashie calls "Black aliveness" (2021).

These two forms of transmutation also help us reveal the power dynamics in ARTs, where the negative transmutation of reproductive labor into capitalist value is being countered by the positive transmutation of reproductive injustice into hope.

While visiting The Frank-Ratchye STUDIO for Creative Inquiry at CMU in August 2024, Mengist got to flip through an archive of subRosa documents and paraphernalia from over the years. One of these was a workbook for a larger project titled *Sex and Gender in the Biotech Century.* In a reflection on the project in their 2002 essay "Stolen Rhetoric: The Appropriation of Choice by the ART Industry," subRosa argues that "(T)he ideology and practice of new eugenic principles that are part of the macro discourse of ARTs has been masked by the micro discourse of individual choice" (subRosa 2002, 143). Many of subRosa's predictions have come to pass — something that is also said of Dorothy Roberts and Gena Corea. And subRosa also looked to alchemy to envision the present moment differently.

Reproductive Mutual Aid

Dorothy Roberts (2011) predicted ARTs scenarios that have come true — largely through conversations with other women and careful observation of clinical and legal trends. Roberts's emphasis centers on Black women's and families' experiences, helping fill some gaps that subRosa also identified in "Stolen Rhetoric" (subRosa 2002). Literature played a part as well. For Roberts, it was *The Handmaid's Tale* that foreboded the present, alongside Gena Corea's predictions that called attention to a historical precedent for the current entanglement of myth, male domination, and biotechnology. Before Roberts's (2011) chapter "Race and the New Biocitizen," Corea's 1985 text also made the prediction that "there will be no great demand for the eggs of a black woman" but that "there may well be a demand for her womb" (Corea 1985, 2). For Roberts, the coupling of IVF and preimplantation genetic diagnosis (PGD) represented a new cautionary tale in 2011. Taking advantage of "reprogenetic" technologies, companies allow "biocitizen parents" to control every step of reproduction (Roberts 2011, 212–213). While women of color with access to the necessary financial resources may be provided opportunities for these reprogenetically perfected babies, low-income women of color (without credit history and health insurance) might actually be discouraged from "having babies at all" (Roberts 2011, 213).

Media coverage and the popular imaginary around ARTs were first visibly racialized as white. Roberts marks the shift of this dynamic from the 1990s to twenty years later, when fertility clinics marketing their services to people of color started depicting smiling black and brown families on their websites (Roberts 2011, 214). Advertising like this may have encouraged those with the means to access IVF, but Roberts notes that governmental welfare programs have not supported funding these options for poor women of color. She goes further to suggest that lawmakers and public health policies may not only continue encouraging poor women of color to have fewer children, but that they may also "pressure them to have genetically screened children": "Unlike IVF, whose primary purpose is to increase fertility, PGD functions to help women avoid starting a pregnancy that entails unwanted genes. The aim of IVF is to produce the birth of a live baby; the aim of PGD and fetal diagnosis is to prevent the birth of certain children" (Roberts 2011, 220)

Does that mean that not all ARTs are created equal? That reprogenetics take IVF outside of its lane and can promote reproductive oppression when misused? This is where it would be helpful to explore some of the nuances in Roberts's argument:

> Genetic technologies by themselves cannot undo or transcend social inequalities. The most advanced and expensive ones will be reserved for the wealthiest people and fall outside the reach of most women of color. The market will privilege a tiny elite among people of color who can afford high-tech reproductive innovations while relegating the vast majority to the most

> intense reproductive surveillance. The application of market logic to childbearing is likely to expand the hiring of poor and working-class women of color for their reproductive labor. Paying these women to gestate fetuses or to produce eggs for genetic research could intensify, even as the very same women are encouraged to use genetic technologies to screen their own children, limiting their reproductive freedom. (Roberts 2011, 222)

What if the technologies are not used in isolation? What if women of color had some reserved for them, and PGD could be refused or not recorded? What if the poor/working class had union protections for their reproductive labor and research participation while receiving additional benefits toward their own reproductive liberation? These questions follow Roberts's own logic: if IVF is for increasing fertility and birthing live babies, then what are the conditions for making that a reality for poor women of color, too, and can it be achieved without coming at the expense of other people's reproductive freedoms? Roberts argues that the Indian surrogacy market represents this exchange, where rich "biocitizen parents" contribute to an illusory vision of reproductive liberation enabled by ARTs that take life from some and sell it to others.

There are valid arguments for why this critical approach may obscure certain women's choices to pursue reproductive labor as dignified work (see, for example, Marwah and Nadimpally 2018), yet the fact remains that the people struggling for the means to access these technologies are not the ones driving the innovations or disseminating the information about these technologies. Although they are not necessarily powerless, they are not nearest to the power nor the profits that these technologies generate. Here, we are especially interested in how collective power functions in these spaces, and how arming people with knowledge draws out the possibilities for both benefits and bigger questions to be developed together.

Cynthia Greenlee provides many updated examples in this regard. Recently in *The Guardian* (2023), she reported that although ARTs are often out of reach for low- and mid-income black women in the U.S., communities have been taking matters of fertility into their own hands through a form of mutual aid. According to Greenlee, the effectiveness of now normalized but still expensive practices of in vitro fertilization (IVF) and intrauterine insemination (IUI) are being supplemented and assisted by Black women's care communities, which freely circulate "information, medications, money, and even donating biological material such as sperm, eggs, and embryos" (Greenlee 2023, n.p.). After talking with over 40 Black women, not unlike Roberts in her early research for *Killing the Black Body* (1997), Greenlee began to see an unprecedented picture. Beyond what Corey, Roberts, and even subRosa all feared, there is more than just harm taking place among Black women navigating the fertility industry. Certainly, there are inequalities, and this point does not deny that: Black women are more likely to experience infertility but less likely to be referred for fertility treatment, less likely to have health insurance that covers IVF or at all, less likely to have access to sperm donors of their race, etc. (Greenlee 2023).

However, black women are also increasingly turning to their own self-defined mutual aid networks, sharing material resources as well as "support that ... they say they rarely receive it in clinical settings" (Greenlee 2023, n.p.). Their strategies include egg sharing along expanded ideas about what family can mean in the context of African American kinship traditions. In this, they perform what Dána-Ain Davis calls "racial reconnaissance" (also cited in Greenlee 2023) in order to scout/identify which fertility clinics will not cause them harm but rather provide racial concordance, and utilizing online spaces to ask medical questions, time their procedures, or even "give, swap or sell" expensive fertility medications (Greenlee 2023, n.p.). One insight Greenlee notes is that Black women and donors prioritize sharing these and other precious reproductive resources with other Black people in order to "contribute to making Black families possible" (Greenlee, 2023, n.p). Here is one story about egg-sharing:

> This past summer, Kailey Townsend called her co-worker Jamila Galloway with a generous offer: she wanted to give her an egg, one of many she'd retrieved and stored for future use. Because Townsend has endometriosis, a condition in which the uterine lining grows outside the organ, and a bleeding disorder related to hemophilia, she decided to act when she had an abundant supply of youthful eggs in case her conditions worsened later. The 27-year-old's egg freezing procedure yielded more than she'd ever need.... For Townsend, offering this part of herself to Galloway was uncomplicated. Her brother had donated sperm to someone, and her family and partner approved. Townsend said that giving an egg to Galloway fits within her expansive notion of family and African American tradition. (Greenlee, 2023, n.p.)

In the Winter of 2024, Mengist was part of a small group of graduate students and faculty hosted by the Simpson Center for the Humanities at the University of Washington, Seattle. As a part of a series of events based around the work of Dorothy Roberts, the group had an opportunity to engage with her directly. When Mengist described Greenlee's egg-sharing story as possibly fitting what we might think of as "reproductive mutual aid" with Dorothy Roberts, she asked "is it truly a radical break? What are the values behind it?" (Roberts and Mengist 2024). This led to many more questions. For example, the values expressed by Kailey Townsend are those of a Black American kinship tradition, but Greenlee makes no clear critique of the conditions that led Townsend to freeze her eggs. Did the extraction of this "abundant supply" arise from the imposition of a scarcity logic, and was Townsend coerced into freezing so many eggs? Who paid for the procedure and who profited from it?

Black American kinship values are set within a broader religious context operating in the fertility industry. Greenlee discusses fertility doulas that promote Afro-diasporic spiritual practices and coach others on how to navigate IVF are incorporating traditional values that facilitate culturally informed decisions and outcomes for Black birthing people (Greenlee, 2023). Similarly, *Fertility for Colored Girls + Guys* is a Black church organization that provides grants, support groups, prayer circles, workshops, and a Facebook Group with over 8,000 members at the time of Greenlee's article (2023). As mentioned earlier in the discussion of alchemical reproduction, Christian ethical interpretations of ARTs about the beginnings of life are framed as "the age-old friction between faith and science" by Laura Ungar and Tiffany Stanley in an article from *The Associated Press* (Ungar and Stanley 2024, n.p.). They report on how families who self-identify as Christian (and all seem to be white in this report) navigate beliefs and decisions about IVF in ways that are in line with our framework of compassionate inquiry. Some Christian couples circumvent hard-line anti-science standpoints that IVF strays too far into the territory of God (which was also an argument made against alchemists) by opting to "adopt" frozen embryos that have been "leftover" (Ungar and Stanley 2024, n.p.). Though motivated by a different set of values than critics of the capitalist impulses in the fertility industry, there is some overlap in the desired results—less pouring of capital into the freezing of more embryos. Additionally, like Kailey Townsend, they believe donating frozen embryos can be a form of kinship (Townsend, quoted in Greenlee 2023, n.p.). Similarly to *Fertility for Colored Girls + Guys*, this process is facilitated by a service called *Snowflakes*:

> Some couples and religious leaders find an answer in embryo adoption, a process that treats embryos like children in need of a home. Snowflakes, a

> division of Nightlight Christian Adoptions, has offered this service to more than 9,000 families since its inception in 1997, with more than 1,170 babies born. Executive Director Elizabeth Button said they got an influx of inquiries after the Alabama ruling. (Ungar and Stanley 2024, n.p.)

One difference between organizations like *Snowflakes* and *Fertility for Colored Girls + Guys* is that *Snowflakes* opposes some of the reproductive options that people seek, while *Fertility for Colored Girls + Guys* supports a broader range of reproductive options, including adoption.

In their open-minded approach, it seems that Black American communities fill some of the gaps that Roberts identified in "Race and the New Biocitizen" (2021). One question that remains is whether these are the decisions made exclusively by Black women without financial means or if black women with wealth also contribute to each other's fertility in these ways. Or perhaps they pursue surrogacy more often, because it is more affordable to them and they might also be more inclined to use PGD. In any case, Cynthia Greenlee has shown that the picture is not as clear or bleak as previously painted, even though there are still complications to engage with. There is indeed a case for reproductive mutual aid here, and potentially an exciting opportunity to enact a *social practice bioart* that makes space for more Black women and birthing people to imagine what else they might do for each other and other communities.

Perhaps, Biocitizen is not how these people think of themselves or their desires—maybe they are consuming these services but are simultaneously operating as bioactivists, bioorganizers, and biocommunities in an uncertain time of a promise of reproductive justice within the circumscription of biocapitalism. This is why work that does not rush to condemnation with regard to choices to pursue ARTs or IVF is important to our own conception of feminist compassionate inquiry (Aristarkhova 2018). Seeing the humanity and dignity in these women's stories is required for a conversation about the historical and future stakes of reproductive technoscience. Compassionate inquiry acknowledges harm that has been done to specific communities to move forward otherwise.

Life with "Abortion Joy"

In this section, we want to acknowledge other examples of reproductive mutual aid around the world. As Tania Kupczak emphasized in Chap. 3 of this book, *The Domain Errors! A Roundtable*, grassroots reproductive mutual aid is the only path to deal with oppressive systems that will never support the lives and "interests of marginalized communities," and provided several international examples:

> Having been wrapped in labor organizing, mutual aid, and pro-abortion work for the last decades, the most salient message I hear over and over again is "We will save us." … We must create our own systems of care, including a rejection of the absolutism of legality, in a framework of repro justice and mutual aid. Embodied knowledge must keep rising as a resistance to patriarchal power dynamics in legislative, medical, and academic spheres. A great example is the Four Thieves Vinegar Collective, an anarchist biohacking group that offers recipes for DIY pharmaceuticals. The use of misoprostol in self-managed abortions owes a debt to a group of Brazilian feminists in the 1980s who began testing medications with pregnancy warnings. Here in the US, we can learn so much from our comrades around the world who have been resisting oligarchical regimes for generations. (Kupczak 77, this volume)

Honoring life that already exists and the agency of those who seek aid includes compassion and joy with which the mutual aid is rendered. Listening to persons themselves, holistically, without separating their physical needs from emotional needs, or social and economic needs, creates a new community. While couples in Alabama, who lost their "potential children" in the "cryogenic nurseries" of an IVF clinic, chose to seek accountability for their loss and grief through the "absolutism of legality" and sued the IVF clinic (which led to the infamous Alabama Supreme Court ruling from 2024), a closer look at Kupczak's example of Brazilian feminist legacy with misoprostol shows that "our own systems" include creation of self-sustaining networks, where "private" loss and grief, or pleasure and joy are finally seen, acknowledged, and shared by other members of community. What Kupczak refers to here, the use of misoprostol, has grown into a system of feminist reproductive mutual aid across South America and beyond. Lea Happ's article "Making Feminist Selves: Reflections on the Biopolitical Working of Misoprostol" (2025) details the affective dimension of reproductive mutual aid in the use of misoprostol based on her own and others' research, enabling us to add this angle to reproductive mutual aid.

In this article, Happ argues that Argentinian feminist communities of support and care among abortion activists and community members have created an environment where persons seeking abortion from these groups are not only supported through the procedure and helped in overcoming negative emotions associated with abortion access (fear, shame, intimidation, poor provider support especially for poorer and marginalized communities of indigenous and trans/queer pregnant persons) – they transition towards pleasure and joy of abortive process as a moment of "clarity and empowerment" through feminist community support:

> From the 1990s, Argentine activists have increasingly relied on misoprostol (McReynolds-Pérez 2014), making illegal abortions safer and enabling new forms of feminist abortion support. In 2009 Lesbians and Feminists for the Decriminalization of Abortion (Lesbianas y Feministas por la Descriminalización del Aborto) launched a phone line sharing instructions for its safe use (Drovetta 2015). Since 2012 Socorristas en Red (feministas y transfeministas que abortamos) (SenRed) has informed and accompanied abortion seekers in and outside the healthcare system (Socorristas en Red n.d.-b). (Happ 2025, 2)

Subjectivities formed through reproductive justice work are intimately connected to the embodiment and decisions about already existing life and enact changes in its treatment. Happ argues that the process itself and the community it has created invented new forms of "feminist identity and subjectivity," which are "counter-hegemonic," as Happ builds on the work with affect theory in relation to abortions by, among others, Nayla Luz Vacarezza, Julia Burton, and Mariela Solana.

Displacing abortion as something with only sharply shaming and traumatic connotations, Happ presents misoprostol as a "biopolitical technology" that enables "biopolitics from below," borrowing the term from Panagiotis Sotiris and Facundo Rocca. Happ theorizes Sotiris and Rocca's vision further towards Argentinian activists' "affective attachments as vehicles for making distinctly feminist selves" (Happ 2025, 2).

In the section aptly titled "Abortion Joy" and highlighted in the title of this section, Happ further emphasizes Ni Una Menos activists' request to imagine new worlds "according to a feminist ethics of life, not of sacrifice, [but] of pleasure and desire" (Menos, quoted in Happ 2025, 17):

> Emotional responses such as pleasure, joy, and desire to abortion have been made possible through feminist activist accompaniment practices (Kimport et al. 2023). As one person sharing her testimony in Estamos Cerca describes, "I remember them [the two activists who accompanied her] and I cry tears of joy, because thanks to the two wonderful people I met,

> I fulfilled my decision and I'm proud of the decision I took" (Socorristas en Red 2020c, 15). Her statement underscores how feminist activist practices made possible through technical expertise about misoprostol and practices of solidarity outlined above enabled her feelings of joy and pride. (Happ 2025, 17)

Happ also focuses on aesthetic choices to drive her point about life and passion as central to this new feminist subjectivity, including the use of the color pink in the SenRed network's website and other visual communication. This process is "life-affirming," as Happ shows through her research on testimonies by those who used the service:

> In Estamos Cerca, various narrators recount the happiness and relief after their abortion, as exemplified by statements such as, "I knew that at some point the embryo would come out and it did. That was it, everything had already happened, I felt tranquil. I couldn't be happier, because it sounds a little rough, but I was happy" (Socorristas en Red 2020b, 8). … In conjunction with providing pregnant people with solidarity networks and the expertise to confront medical authorities, a "discourse of celebratory abortion" (McReynolds-Pérez 2014, 157) provides a sense of agency over their own bodies otherwise denied. Within conventional perceptions of abortion as "always already…something 'wrong'" (Sutton and Vacarezza 2020, 734), celebratory discourses introduce pleasure as a valid priority into the debate. …. One person states that "I confirm that motherhood must be desired, or it must not be" (Socorristas en Red 2020d, 13), while another says "my uterus gestates the poetry of freedom" (Socorristas en Red 2020b, 10). (Happ 2025, 17–18)

Similar to the ART industry, around the world, wealthy women are rarely as affected by reproductive restrictions and injustice as poorer women, who also do not have as much privacy in decisions about their own bodies and lack access to private clinics. These feminist networks overcome this situation by creating safer and accompanied experiences that continue to transform and constitute new, joyful subjectivities. These networks have also been instrumental in changing reproductive justice laws, empowering more people to experience this freedom and reproductive biomedical support.

Within this context of misoprostol's mutual aid activism, it would be useful to mention another group that was part of the tactical media community at the time when subRosa started its work. This other group of reproductive activists crossed paths with subRosa members at the *Next Five Minutes Festivals of Tactical Media* in Europe in the early 2000s. The group is called

"Women on Waves." Established around 1999–2000, they embraced the use of pills to end pregnancies as a form of reproductive justice activism in countries, especially in Europe, where reproductive choice was banned or inaccessible. Started by Dr. Rebecca Gomperts, a Dutch activist, medical doctor, and artist, the *Women on Waves* group procured a ship that could legally enter international waters off a 12-mile radius to the land, which they did, such as in Ireland or Portugal, to attempt to provide legal terminations, in accordance with the Dutch law. The ship carried a mobile abortion clinic that was approved by the Dutch Ministry of Health for these purposes.

To design and create the mobile clinic pod on the ship, Gomperts partnered with an architectural design bureau, Atelier Van Lieshout, and Joep Van Lieshout created it for *Women on Waves.* They called the object "A-Portable," and in addition to being registered as a mobile clinic, it got a status as a work of art and design. This structure – a container decorated and medically supplied – has been exhibited many times since then, most notably at a Venice Biennale exhibition in 2001. For Gomperts and Van Lieshout, this project was always part of their art and design practice. For Gomperts, it was also about biomedical practice and activism related to decisions about life and reproductive freedom (see more on the *Women of Waves* website: https://www.womenonwaves.org/en/). Today, *Women on Waves* continues to send pills by mail (also by drones). Their practice of challenging governments uses direct action and has been extensively written about (see, for example, Jennifer Bisgaier 2019) and starred in Diane Whitten's award-winning documentary film *Vessel* (2014), testifying to their ongoing cultural, social, legal, biomedical, and aesthetic impact. Gomperts also represents, arguably, a very different kind of tactical biopolitical practitioner, who truly crosses the boundaries of "strong" and "weak," "below" and "above," in the formulations of biopolitics, blending the role of science, medicine, art, and citizen activism in reproductive mutual aid. These examples provide similar reproductive services in strikingly different forms of what support and mutual aid could look like.

subRosa emphasized that their cyberfeminist work "interrogates the intersecting ideologies and practices that serve to normalize and naturalize ARTs, exposing their historical connections to eugenics and colonial ideologies" (2002, 148). In addition to interrogation and exposition, we want to build compassionate inquiry about the use and development of new ARTs while keeping this disempowering history in mind. Those who have been excluded from access to ARTs or other medical scientific applications know very well the circumstances of their own exclusion and are often wary of entrusting their bodies to mainstream biomedicine and technoscience—for good reasons. That is why interrogation and exposition need to happen, especially on behalf of those *who have access* and who benefit from that access.

For those who do not have access to start with, interrogation and exposition can only get them so far. In response to a question by a physician whose patients struggle with decisions about ARTs at a panel about the global history of abortion care and politics since the 1960s, Alys Weinbaum quite generously offered that although the need for a critical historicization of reproductive racial capitalism's technologies is more and more pressing in these post-Roe times in the United States, an appreciation for the pragmatics of people's everyday decision-making about those technologies is also needed (Weinbaum 2024, Morgan and Weinbaum 2024). In alignment with Cynthia Greenlee's reporting and theorizing, this is what our framing of compassionate inquiry adds to previous cyberfeminist interventions as well as to Dorothy Roberts's critical scholarship. Building on reproductive mutual aid in the use of ARTs and abortion access imagines other types of reproductive justice. It also provides new tools for mutual support.

Roberts draws the framing of "biological citizenship" from Nikolas Rose's work on Michel Foucault and from Paul Rabinow (Roberts 2011, 205–206). Though there is no space here to elaborate in detail on this point, taken together with Zieglerin's subversive vision of augmenting her own reproductive body and the story of *DIY Homunculi*, reproductive mutual aid practices are a valiant repurposing of ARTs toward liberatory ends—though not necessarily purely virtuous, as they are wrought with power imbalances, conflicting desires, and contradictions. They are "alchemies of birth," transmuting a circumstance of inequality into something more dignified and diverse. And they connect us to subRosa's cyberfeminist tactical biopolitics.

Indecision Trees

In this section, we come back to the creative worldbuilding of biopolitical cyberfeminist art to get inspired by its refusal of disempowering options. subRosa's "Sex and Gender in the Biotech Century" was first performed in November 2000 at the *Digital Secrets* Conference at Arizona State University, and then once again, retitled as "The Sex and Gender Education Show" for *Hardware, Software, Wetware, and Women* conference at St. Mary's College of Maryland, in December 2002. subRosa collaborated with Michelle M. Wright (also co-editor of *Domain Errors!* and a contributor to Chap. 3 of this book), who performed as "Dr. Eugenica Invitra, Secretary of Federal Eugenics Administration." The work is described on subRosa's website as "the first public performance" by the group, "that employed time-honored, low-tech teaching methods to inspire critical thinking and knowledge sharing about the new Assisted Reproductive Technologies. Just coming into wide use at the time, we looked at ARTs and their effects on female sexuality, reproductive choice, eugenics, and gender in the Biotech Century." (subRosa, at http://cyberfeminism.net/category/work/performances/) In line with SymbioticA (Zurr and Catts 2023), OBN, Critical Art Ensemble, and Yes Men performances, subRosa members used an aesthetic of an official agency, scientific, corporate, or governmental, to interface with the public. At the event itself, subRosa members:

> posed as corporate and government agency representatives, registering class participants and assigning them to one of 5 groups named for the protein bases of DNA (T, A, G, & C), or a fifth, "dud" group. Following an illustrated crash course on reproductive genetics and ART methods, participants were given a workbook with a 'reproductive choice' form to complete. In a hilarious final "repro-tech mixer," participants mingled to find reproductive partner(s) matching their preferences to negotiate "making a perfect baby the biotech way." A spontaneous break-away group of Luddites desiring to have babies the old-fashioned sexy way highlighted this performance. (subRosa, at http://cyberfeminism.net/category/work/performances/)

The setting of a tongue-in-cheek, familiar science education set up provided performers with a cover of legitimacy (white medical/science laboratory gowns, science-speak on the Workbook handout) while also allowing for a less rigid reading of educational games. Some elements were supposed to obscure the work as an art project, while other aesthetic elements pushed the audience to confront the most critical and controversial aspects of reproductive biotechnology (outlandish names of groups, absurdist use of the logic and language of eugenics). These techniques enabled subRosa to bring attention to the way in which sex and gender continued to be employed in normalizing certain reproductive choices and labor associated with them, while also pointing out how those norms have been changing. Especially, the focus was on "new options" in reproductive biotechnology, when tissues become separated from the bodies that they are part of and manipulated outside in a petri dish, or a bio-incubator, or in another person – a surrogate, for example, to achieve the ends of a certain "look" of a child.

What happens to gender, sex, race, body/ability—class in the use of ARTs? These new "options" are where "eugenics" comes in and finds itself anew. In the accompanying publications, subRosa refers to and quotes Richard Lewontin, a geneticist who has been sharply critical of genetic determinism and essentialism. In the interview for the *Tactical Biopolitics* book quoted also in Chapter 1, Gwen D'Arcangelis, Beatriz da Costa and Kavita Philip (2008) asked Lewontin about the history of eugenics and how it corresponds to current interest in human genetic selection. For Lewontin, early development of genetics is intrinsically linked to eugenics, because of the belief that genes—and hence, birth—determine the whole organism and its development:

> Almost all famous geneticists were one kind of racist or another, even those who were antiracist, so to speak. Look, a famous geneticist like Fisher, who founded population genetics, he was a real racist. Most British geneticists were racists. Even people like H. J. Muller, who politically started out as a Marxist and who would not be called politically conservative, nevertheless believed that genes determined just about everything, and he was a very strong eugenicist. So, eugenics was a very important part of genetics for a long, long time. . . . It's been replaced by a medical predictive form of genetics, which is where they look at your genes and say, "Well, look, if you have a kid, it's likely that your kid will blah-blah . . . so be careful." . . . [M]y guess is the majority of geneticists, of working geneticists, believe that genetic differences are pretty important in determining whether you have high IQ. I bet they wouldn't come out and say it, but I still have a feeling that they do. (Lewontin, quoted in D'Arcangelis, da Costa, and Philip 2008, 10–11)

Eugenics has been critiqued in many of subRosa's works. In *Expo EmmaGenics*, a satire on American genetic determinism and early ART industries (see Chap. 6 for details), promised "designer babies" and other genetically engineered biomedical enhancement products, offering them as a lifestyle choice of "menu" options provided by new biotechnologies. One form of this "menu" is a decision tree workbook activity. In *Expo EmmaGenics*, various new biotech options were presented as an "expo," where the audience members, as customers, could sample innovations, such as *LuproBar*, that assist in genetic rejuvenation. There is a video that documents the expo and frames it from the point of view of the American reproductive and genetic industry, and it is available at http://cyberfeminism.net/category/work/publications/video/. Here, we again observe bio-art's performative tropes, such as the heavy use of scientific terms, laboratory equipment, posters, laboratory table setups of biotech academic and industrial conferences, and trade shows. Hence, the title of "biotech expo" and the company's name to fit the style: *EmmaGenics* mimics many biotechnological start-up names that also use lettering with irregular capitalization.

In the *Sex and Gender in the Biotech Century* performance, the audience members were "sorted out": swabbed for genetic material, registered, and divided into several groups. The brochure that accompanied the performance, the Workbook, provided a lot of information and presented this "new eugenics" language as an already-here, happening part of science and ART industries. The audience members were asked to choose various options for their offspring:

In addition to this list of choices, the workbook included this page with "Ten Ways to Enhance Yourself Genetically." The options here list the techniques that are widely practiced today in the biomedical industry, such as "pre-implantation genetic diagnosis through embryo biopsy." Other suggestions mixed with them make the critical point of new eugenic tendencies in the ART research when recommending "parental eugenics and eugenic sacrifice when necessary":

As with many other predictions, this desire to genetically (which could also be termed in Lewontin's terms "eugenically") enhance one's offspring has come true in the global fertility chain. For example, some prospective parents and clinics in China illegally procure donor eggs from women in Eastern European countries as the parents seek specific "appearance and intelligence" traits to achieve a desirable result of having a "fairer" or more "Eurasian" looking child (e.g., see Weis 2021). Taking us back to the beginning of this chapter, it is important to ask whether children who are "made" through this process would also be as thrilled to be "enhanced" in this way.

As subRosa appropriated the form of a decision tree widely used in the ART industries, there is still a lack of scrutiny of its use. We decided to investigate its history to also understand how subRosa borrowed this tactic from the biomedical industry and how it compares to its use in other spheres.

Decision trees have been used in research on ARTs for decades (Jurisica et al. 1998; Guh et al. 2011; Fu et al. 2022; Villani et al. 2022). As a core element of algorithmic induction systems, much of the literature has to do

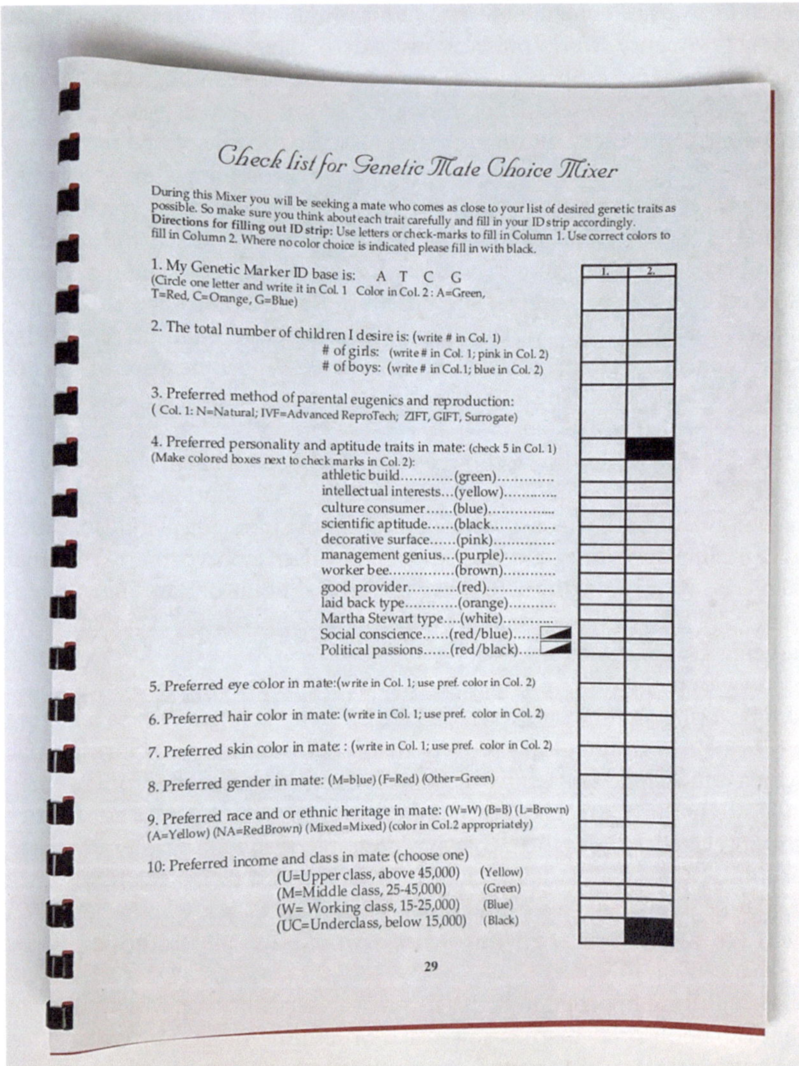
Check list for Genetic Mate Choice Mixer

During this Mixer you will be seeking a mate who comes as close to your list of desired genetic traits as possible. So make sure you think about each trait carefully and fill in your ID strip accordingly.
Directions for filling out ID strip: Use letters or check-marks to fill in Column 1. Use correct colors to fill in Column 2. Where no color choice is indicated please fill in with black.

1. 2.

1. My Genetic Marker ID base is: A T C G
(Circle one letter and write it in Col. 1 Color in Col. 2 : A=Green, T=Red, C=Orange, G=Blue)

2. The total number of children I desire is: (write # in Col. 1)
# of girls: (write # in Col. 1; pink in Col. 2)
# of boys: (write # in Col.1; blue in Col. 2)

3. Preferred method of parental eugenics and reproduction:
(Col. 1: N=Natural; IVF=Advanced ReproTech; ZIFT, GIFT, Surrogate)

4. Preferred personality and aptitude traits in mate: (check 5 in Col. 1)
(Make colored boxes next to check marks in Col. 2):
athletic build............(green)............
intellectual interests...(yellow)............
culture consumer......(blue)...............
scientific aptitude......(black..............
decorative surface......(pink).............
management genius...(purple)...........
worker bee..............(brown)............
good provider...........(red)...............
laid back type...........(orange)..........
Martha Stewart type....(white)...........
Social conscience......(red/blue)......
Political passions......(red/black).....

5. Preferred eye color in mate:(write in Col. 1; use pref. color in Col. 2)

6. Preferred hair color in mate: (write in Col. 1; use pref. color in Col. 2)

7. Preferred skin color in mate : (write in Col. 1; use pref. color in Col. 2)

8. Preferred gender in mate: (M=blue) (F=Red) (Other=Green)

9. Preferred race and or ethnic heritage in mate: (W=W) (B=B) (L=Brown) (A=Yellow) (NA=RedBrown) (Mixed=Mixed) (color in Col.2 appropriately)

10: Preferred income and class in mate (choose one)
(U=Upper class, above 45,000) (Yellow)
(M=Middle class, 25-45,000) (Green)
(W= Working class, 15-25,000) (Blue)
(UC=Underclass, below 15,000) (Black)

29

Fig. 4.9 A page from the comb-bound *Sex & Gender Education* workbook depicting the "Checklist for Genetic Mate Choice Mixer." This page was filled in and used by participants during an interactive activity at the end of the performance (subRosa 2000, 2002).

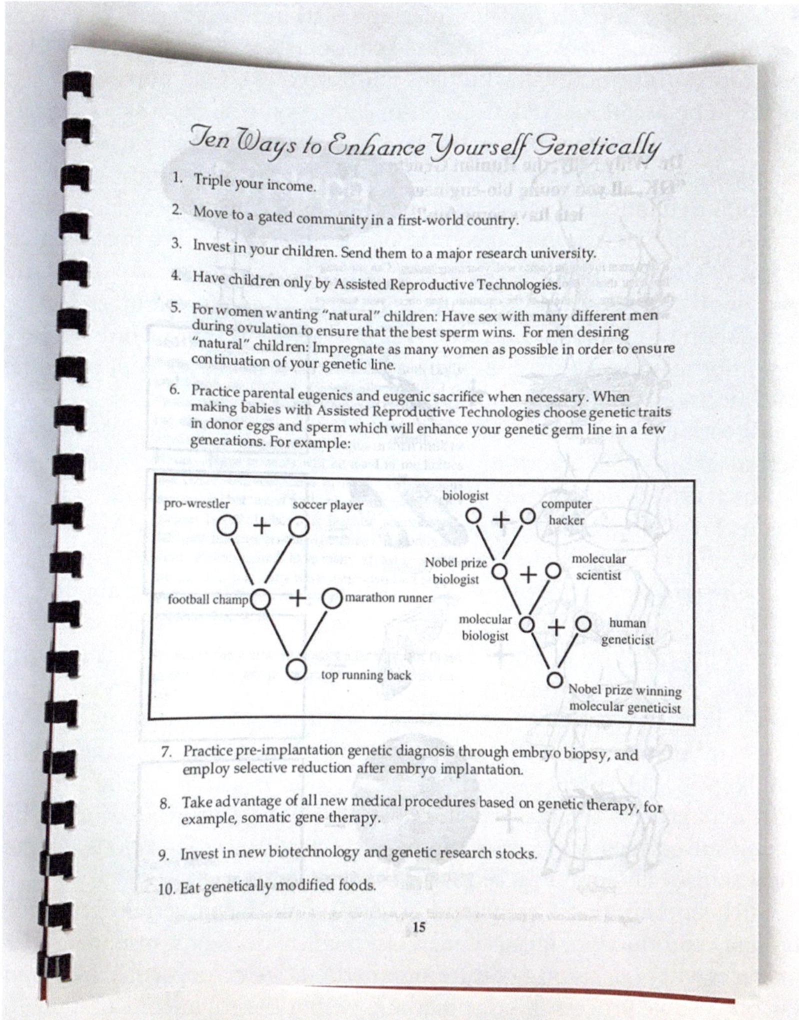

Fig. 4.10 A page from the comb-bound *Sex & Gender Education* workbook depicting "Ten Ways to Enhance Yourself Genetically." The workbook was used by participants throughout the performance and they were allowed to keep it as a memento and resource (subRosa 2000, 2002).

with predictive models that use machine learning in different ways with the ultimate goal of increasing the likelihood of successful IVF and IUI procedures (historically, for humans and for cows). One borrowing that seems to be useful for ART decision-making, going all the way back to its early origins in economics, is the layering of financial information onto each leaf/node of the decision tree. ARTs are expensive, and families, in addition to biomedical parameters, are constrained by many economic factors (e.g., how much insurance may or may not cover). We imagine that being able to see the intricacies of how much it may cost to take one route over another could help with comprehension, communication, and planning among individuals, partners, and providers. In other words, monetary information matters, and being able to see it alongside your potential ART decisions could help with making informed choices.

Historically, decision trees were first applied in the business world before taking off in algorithmic technologies like machine learning models. In a 1964 article in *Harvard Business Review*, John F. Magee (1964) credits French economist, engineer, and mathematician Pierre Massé with the coining of the term in *Optimal Investment Decisions: Rules for Action and Criteria for Choice* two years earlier (Massé 1962; interestingly, the subtitle is reminiscent of the politics of choice critiqued by subRosa). Since then, it has taken off as a decisional support tool in other fields. A relevant example and helpful overview for decision trees in medicine was provided by Vili Podgorelec, Peter Kokol, Bruno Stiglic, and Ivan Rozman (2002), which by that time had already been used in medicine since the 1980s (2002, 458). Here, they describe how decision trees can help healthcare providers process the massive amounts of new data in front of them by taking solved medical cases as inputs and outputting suggestions for the most statistically sound decisions for the current case.

Such optimizations, as helpful as they could be in certain contexts, present reproduction, financial markets, medical decisions, insurance (risk management) decisions—future-oriented, largely unpredictable and uncontrollable processes—as something within the parameters of human, and especially, expert, control. Within scientific authority, gender and sex *seem* even more fluid than in the idea of their "social construction," where machines, humans, animals, and plants potentially co-evolve in the future tense based on human genetic manipulation (CRISPR, short for Clustered Regularly Interspaced Short Palindromic Repeats).

There has recently been a clear show of support for assisted reproductive technologies (ARTs) in the U.S. since the Alabama Supreme Court ruled that frozen embryos are "unborn children" on February 16, 2024 (The Associated Press 2024.) However, abortion access is under threat, and many critical resources have shifted to protecting what abortion access remains. This might mean that fewer questions are being asked about the current state and future of ARTs in a time when much is still unknown about people's experiences. We propose an open-ended exploration of different ethical and aesthetic questions about ARTs through the idea of *Indecision Trees.* Our hope is that, like subRosa has done so many times, our *Indecision Trees* can become a starting point for future creative and educational resources.

Within this broad context, we propose building further out from this speculative world. We have begun to envision a subRosa-inspired practice where we, as bioartists, would provide facilitation materials that could engage participants to imagine what beautiful children could be created if the only limitation on access to ARTs is that they must incorporate and actually depend on plants. Rather than reproducing how contemporary designer babies are marketed toward privileging "superior" human traits, we hope to bring *DIY Homunculi* to life by shifting attention to plants and ensuring that any queer or trans coupling is possible. Our cyberfeminist appropriation of decision trees includes plant-based genetic material as a means of increasing the likelihood of a successful birth. In the spirit of *DIY Homunculi*, these *Indecision Trees* would embody the *pARTs* while also explaining all the different "family-building" options available in the real world.

We also hope that some of the pedagogical methods of subRosa's collective arts and education practice emerge anew in this unique moment, complementing previous interactive engagements where subRosa members took participants through an imaginary ART process. Our future *Indecision Trees* could playfully but earnestly envision a twenty-second century beyond the limits of biotechnical domination. We wonder what meanings could be made through the hospitality to/of plants, and how we as feminists can approach ARTs to imagine freely.

Magical Lives

We want to conclude our vision of Magical Lives by circling back to topics around reproductive labor, justice, and worldbuilding. One inspirational quote from subRosa towards this end goal comes from "Beyond Same-Sex

Marriage Statement: A New Strategic Vision for All Our Families & Relationships," in which activists demand other types of "magic":

> (1) legal recognition for a wide range of relationships, households, and families—regardless of kinship or conjugal status; (2) access for all, regardless of marital or citizenship status, to vital government support programs, including but not limited to health care, housing, Social Security and pension plans, disaster recovery assistance, unemployment insurance, and welfare assistance; (3) separation of church and state in all matters, including regulation and recognition of relationships, households, and families; (4) freedom from state regulation of sexual lives and gender choices, identities and expression. ("Beyond Marriage," quoted in subRosa 2008, 241–242)

We also agree with Fernández that "[w]e must think of reproductive politics as larger than biological reproduction," to follow Laura Briggs' call "that all politics is reproductive politics," where "immigration, living wages, education, housing, health, child and elder care, are all part of reproduction. To these we could add a healthy environment" (Fernández 77, this volume).

Life is Magical in two ways for us here: first, it is magical because we don't know how it works, no matter how many decision trees and probability algorithms we train, and how large the datasets become. One might miss that point as probabilities, fields, and other indeterministic crossovers of various potentialities are fast becoming the primary language of technoscience in various fields that now must lend themselves to computation and Artificial Intelligence (AI). However, in this very moment, just as we invest in computing *everything* inside and outside of ourselves, we get *less* precision, *more* approximation, and are asked to settle on a *crapshoot version of reality*. But is this news to us? Is this the big "discovery" of our times? No. What we are asked to do instead is to pay for the consequences of someone else's decision trees. We are asked to exercise pure "trust" in them as we are constantly reminded by disclaimers, "the past performance does not predict future results." Profits find their private pockets of specific "humans," while losses from mistakes and failures are anonymized to be spread around all of us to avoid corporate and governmental responsibility of ownership. We, when faced with the old-fashioned "bullshit," are gaslighted by industry "experts" who want us to call these algorithmic failures and their built-in biases an AI "hallucination" an intermediary step towards the goal. In this process, we are supposed to always "lose."

No. We already have a space for alchemy, with its allowance for magic and chance, because it has always been part of our imagination, and we own it.

Second, Life is Magical, because it *just happens*. For example, we all come out of *other* people, irrespective of our decisions. But life's maintenance is only magical when it is done by someone else. Labor that goes into growing life, sustaining life, and maintaining life is only mentioned in the reproductive biotech industry when they bill you for *their labor and their machines' operating costs*, but otherwise, its value is rendered invisible while being still, well, extracted. Being in life, living, being part of something so much larger than oneself—from the past, ancient, and the future, unknown, in the present, with others, is also about death and the mystery of metamorphosis and transformation. It is also about knowing that life is beyond any control or authority. To have precise answers and rules of life is to believe in more fiction than is contained in all alchemies. We have long been travelling companions with probabilities and impermanence that probability theory and quantum computing have only recently started to respect and learn from. Live with the mystery of "I don't know" and don't call your lies "hallucinations." The *Yes Species* sings about that: the mystery of everyday breath denied or suppressed, now liberated (also see Federici 2004, who inspired subRosa's art conceptualization, and subRosa 2003).

Our concept of *Indecision Trees* looks into the future where reproductive mutual aid is celebrated and supported, and bioart and tactical biopolitics include reproductive labor and justice as part of their interventions and solidarities. If those who breathe and give life and nurture it are "weak" (in de Certeau's terms as discussed in Chap. 1) and have to use various short-term tactics rather than long-term strategies, it is because they have been systematically and violently weakened and disempowered by those who appropriate their strength and energy, with the use of law, military, brutal physical force of sterilization or sexual violence, but also, who seek to capture their imagination through cultural institutions and narratives. In reappropriating imagination, we assert the strength of decision-making as a collective open-ended act, and not just an individually pre-determined biomedical choice. This collective act has nourished magical life for many centuries.

As participants in the *Sex and Gender in the Biotech Century* mixer were assigned their DNA base pairing (either A, C, G, or T), with an added "Dud" group of "losers" "unfit" for reproduction, rebels appeared who refused the given options.

The Luddites group disobeyed during the event and did not follow the instructions of this brave new world of reproductive biotech options of

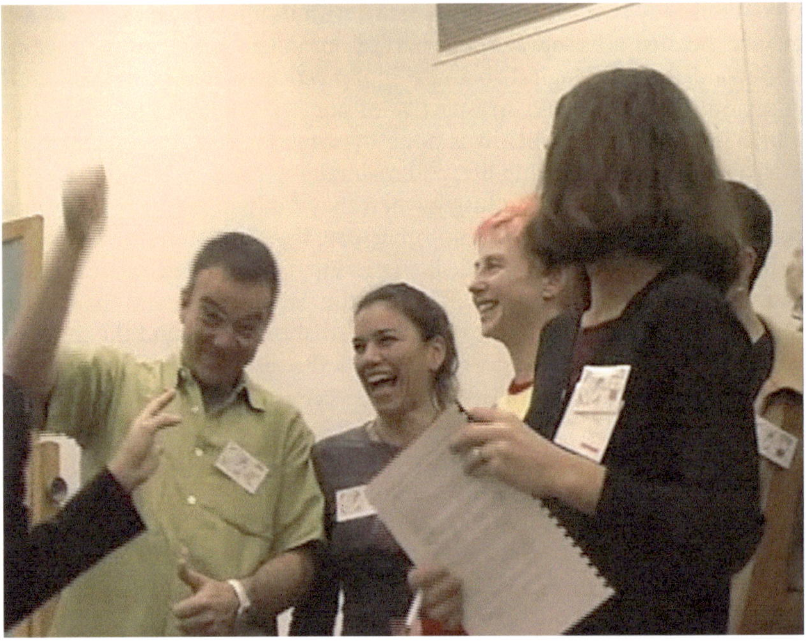

Fig. 4.11 Performance stills from the *Sex & Gender Education* show (2000). At the end of the performance, participants were invited to share the results of a collaborative "mixer" during which they found others in the room with whom to collaborate on commissioning "the perfect baby" according to their self-described compatible desires and traits. DUD Group. "Digital Secrets: New Collaborations in Visual Art and Technology" conference, Tempe, Arizona (video still from sub-Rosa 2000).

the decision trees and charted their own course based on their own manifesto. The rebels look happy, joyful, and defiant as they stand up and openly refuse (Fig. 4.11). Their hands are up in fists, and they stand next to each other in solidarity (Fig. 4.12). Their objections to following the rules of the performance game and "mate" according to their assigned roles and genetic profiles have been welcomed by subRosa members because such a refusal of pre-given "options" was embedded in their own cyberfeminist practice. This refusal, certainly practiced within the "safe" environment of an art performance, was an experiment in defiance, a test-run, so to speak, for real life. After all, this is what subRosa's bioart education show wanted

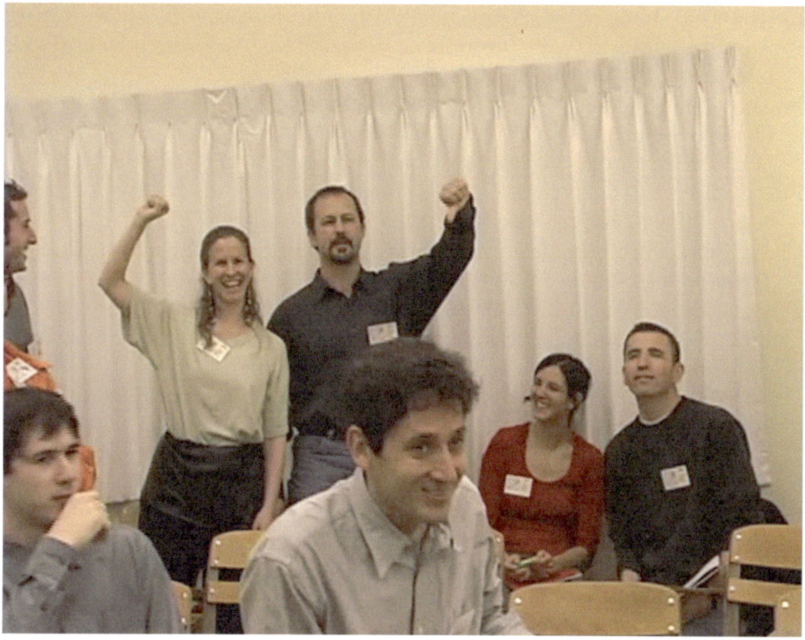

Fig. 4.12 Performance stills from the *Sex & Gender Education* show (2000). At the end of the performance, participants were invited to share the results of a collaborative "mixer" during which they found others in the room with whom to collaborate on commissioning "the perfect baby" according to their self-described compatible desires and traits. Some groups rebelled and stated that they preferred to reproduce "the old-fashioned way." This is the Luddute Group. "Digital Secrets: New Collaborations in Visual Art and Technology" conference, Tempe, Arizona (video still from subRosa 2000).

to prompt, a questioning of the biomedical reality as the inevitability of presented options, choices, and decisions. The Luddites stayed and participated by opting to create something of their own and in the process completed the work.

References

Aristarkhova, Irina. 2012. *Hospitality of the Matrix: Philosophy, Biomedicine, and Culture*. Columbia University Press.

Aristarkhova, Irina. 2018. "Suspending Feminist Judgment." *Boston Review*, August 10. https://www.bostonreview.net/forum_response/irina-aristarkhova-suspending-feminist-judgment/

Bisgaier, Jennifer. 2019. "Navigating Rough Seas: Women on Waves' Legal Options for Overcoming Resistant States." *Chicago Journal of International Law*, 20 (1): Article 3. https://chicagounbound.uchicago.edu/cjil/vol20/iss1/3

Brock, André. 2016. "Critical Technocultural Discourse Analysis." *New Media & Society* 20 (11). https://doi.org/10.1177/1461444816677532

Corea, Gena. 1985. *The Mother Machine: Reproductive Technologies from Artificial Insemination to Artificial Wombs.* 1st ed. Harper & Row.

D'Arcangelis, Gwen, Beatriz da Costa, and Kavita Philip. 2008. "Interview with Richard Lewontin." In *Tactical Biopolitics: Art, Activism, and Technoscience*, edited by Beatriz da Costa and Kavita Philip. MIT Press.

Federici, Silvia. 2004. *Caliban and the Witch: Women, the Body, and Primitive Accumulation.* Autonomedia.

Fertility for Colored Girls + Guys. n.d. https://www.fertilityforcoloredgirls.org. Accessed July 2025.

Fu, Kaiyou, Yanrui Li, Houyi Lv, Wei Wu, Jianyuan Song, and Jian Xu. 2022. "Development of a Model Predicting the Outcome of In Vitro Fertilization Cycles by a Robust Decision Tree Method." *Frontiers in Endocrinology* 13: 877518. https://doi.org/10.3389/fendo.2022.877518

Gearhart, Dakota. 2023. "Episode 3: Garden Spawn." *Life Touching Life.* https://www.youtube.com/watch?v=wzk6WUstp3Y

Greenlee, Cynthia R. 2023. " 'I Couldn't Believe This Act of Kindness': How Black Women Trying to Get Pregnant Create Their Own Healthcare Networks." *The Guardian*, December 11. https://www.theguardian.com/us-news/2023/dec/11/black-women-infertility-mutual-aid-healthcare-networks-medication-sharing-egg-sperm-donations

Guh, Ruey-Shiang, Tsung-Chieh Jackson Wu, and Shao-Ping Weng. 2011. "Integrating Genetic Algorithm and Decision Tree Learning for Assistance in Predicting in Vitro Fertilization Outcomes." *Expert Systems with Applications* 38 (4): 4437–4449. https://doi.org/10.1016/j.eswa.2010.09.112

Happ, Lea. 2025. "Making Feminist Selves: Reflections on the Biopolitical Working of Misoprostol in Argentine Abortion Activism." *Catalyst: Feminism, Theory, Technoscience* 11 (1): 1-27. https://doi.org/10.28968/cftt.v11i1.40522

Jackson, Zakiyyah Iman. 2021. "Against Criticism: Notes on Decipherment and the Force of Things." In *No Humans Involved*, edited by Erin Christovale. DelMonico Books/Hammer Museum. https://delmonicobooks.com/book/no-humans-involved/

Jurisica, Igor, John Mylopoulos, Janice Glasgow, Heather Shapiro, and Robert F Casper. 1998. "Case-Based Reasoning in IVF: Prediction and Knowledge

Mining." *Artificial Intelligence in Medicine* 12 (1): 1–24. https://doi.org/10.1016/S0933-3657(97)00037-7

King, Tiffany Lethabo. 2019. *The Black Shoals: Offshore Formations of Black and Native Studies*. Duke University Press.

Magee, John F. 1964. "Decision Trees for Decision Making." *Harvard Business Review*.

Marwah, Vrinda, and Sarojini Nadimpally. 2018. "Surrogacy and Social Movements in India: Towards a Collective Conversation." In *Cross-Cultural Comparisons on Surrogacy and Egg Donation: Interdisciplinary Perspectives from India, Germany and Israel*, edited by Sayani Mitra, Silke Schicktanz, and Tulsi Patel. Springer International Publishing. https://doi.org/10.1007/978-3-319-78670-4_10

Massé, Pierre. 1962. *Optimal Investment Decisions: Rules for Action and Criteria for Choice*. Prentice-Hall.

Mengist, Nathanael Elias. 2023. "Out of the Flask, Into the Womb: Alchemical Reproductive Politics in Early Modern Visions of Artificial Life." In *The Anthropocene Laboratory*, edited by Fröydi Laszlo and Anna Risell. Gothenburg Publishing.

Mengist, Nathanael Elias. 2025. "Alchemical Reproductive Technologies and Black Birth." Paper presented at *(Re)Imagining Reproductive Justice 2* panel, chaired by Monica Casper. 50th Annual Meeting of the Society for Social Studies of Science (4S), Seattle, WA, September 3–6.

Morgan, Jennifer, and Alys Weinbaum. 2024. "Introduction: Reproductive Racial Capitalism." *History of the Present* 14 (1): 1–19. https://doi.org/10.1215/21599785-10898330

Newman, William R. 2020. "Bad Chemistry: Basilisks and Women in Paracelsus and Pseudo-Paracelsus." *Ambix* 67 (1): 30–46. https://doi.org/10.1080/00026980.2020.1720341

Newman, William R. 2005. *Promethean Ambitions: Alchemy and the Quest to Perfect Nature*. University of Chicago Press.

Nummedal, Tara. 2021. "Alchemical Bodies: Discursive and Material Visions." *Early Modern Women* 15 (2): 121–32. https://doi.org/10.1353/emw.2021.0020

Nummedal, Tara. E. 2001. "Alchemical Reproduction and the Career of Anna Maria Zieglerin." *Ambix* 48 (2): 56–68. https://doi.org/10.1179/amb.2001.48.2.56

Podgorelec, Vili, Peter Kokol, Bruno Stiglic, and Ivan Rozman. 2002. "Decision Trees: An Overview and Their Use in Medicine." *Journal of Medical Systems* 26 (5): 445–63. https://doi.org/10.1023/A:1016409317640

Practicing Refusal Collective. 2021. "Enclosures: Blackness and Transmutation." *The Sojourner Project*. https://www.thesojournerproject.org/sessions/enclosures-blackness-and-transmutation/

Quashie, Kevin. 2021. *Black Aliveness, or A Poetics of Being*. Duke University Press.

Roberts, Dorothy, and Nathanael Elias Mengist. 2024. Public Discussion, January 11. University of Washington, Seattle.

Roberts, Dorothy. 2011. *Fatal Invention: How Science, Politics, and Big Business Re-Create Race in the Twenty-First Century*. The New Press.

Roberts, Dorothy. 1997. *Killing the Black Body*. Pantheon Books.

subRosa. 2002. "Stolen Rhetoric: The Appropriation of Choice by Art Industries." In *Domain Errors! Cyberfeminist Practices*, edited by María Fernández, Faith Wilding, and Michelle M. Wright. Autonomedia.

subRosa: A cyberfemininist art collective. 2003. "International Markets of Flesh (IMF)." http://cyberfeminism.net/imf/

subRosa. 2005. "Useless Gender: An Immodest Proposal For Radical Justice." In *Yes Species*, by subRosa and James Pei-Mun (Wu) Tsang. Sabrosa Books.

subRosa, and Pei-Mun James (Wu) Tsang. 2005. *Yes Species*. Sabrosa Books.

subRosa. 2008. "Common Knowledge and Political Love." In *Tactical Biopolitics: Art, Activism, and Technoscience*, edited by Beatriz da Costa and Kavita Philip. The MIT Press.

subRosa and Dr. Michelle M. Wright. 2000. "Sex and Gender in the Biotech Century." Digital video, 5 minutes and 15 seconds. Private collection. Watched July 2025.

Sudan, Rajani. 2004. "Mud, Mortar, and Other Technologies of Empire." *The Eighteenth Century* 45 (2): 147–69.

Sudan, Rajani. 2016. *The Alchemy of Empire: Abject Materials and the Technologies of Colonialism*. Fordham University Press.

The Associated Press. 2024. "Alabama Supreme Court Rules Frozen Embryos Are 'children' under State Law." *NPR*, February 20. https://www.npr.org/2024/02/20/1232815486/alabama-supreme-court-frozen-embryos

Tsang, James Pei-Mun (Wu). 2005. "Vocalization in an Ethical Matrix." In *Yes Species*, by subRosa and James Pei-Mun (Wu) Tsang. Sabrosa Books.

Ungar, Laura, and Tiffany Stanley. 2024. "A Court Ruled Embryos Are Children. These Christian Couples Agree yet Wrestle with IVF Choices." *The Associated Press*, June 5. https://apnews.com/article/ivf-embryos-christian-evangelicals-alabama-dce11430993cf7e1047987fa3fb06ace

Vertommen, Sigrid, Bronwyn Parry, and Michal Nahman. 2022. "Introduction: Global Fertility Chains and the Colonial Present of Assisted Reproductive Technologies." *Catalyst: Feminism, Theory, Technoscience* 8 (1) : 1–17. https://doi.org/10.28968/cftt.v8i1.37920

Villani, Maria Teresa, Daria Morini, Giorgia Spaggiari, Chiara Furini, Beatrice Melli, Alessia Nicoli, and Francesca Iannotti. 2022. "The (Decision) Tree of Fertility: An Innovative Decision-Making Algorithm in Assisted Reproduction Technique." *Journal of Assisted Reproduction and Genetics* 39 (2): 395–408. https://doi.org/10.1007/s10815-021-02353-4

Weinbaum, Alys Eve. 2024. Public Discussion, April 26. University of Washington, Seattle.

Weis, Christina. 2021. "Changing Fertility Landscapes: Exploring the Reproductive Routes and Choices of Fertility Patients from China for Assisted Reproduction in Russia." *Asian Bioethics Review* 13 (1): 7–22. https://doi.org/10.1007/s41649-020-00156-w

Whitten, Diana. 2014. *Vessel.* Documentary Film. Sovereignty Productions in association with: Fork Films, Impact Partners, Chicken and Egg Pictures.

Wynter, Sylvia. 2003. "Unsettling the Coloniality of Being/Power/Truth/Freedom: Towards the Human, After Man, Its Overrepresentation—An Argument." *CR: The New Centennial Review* 3 (3): 257–337. https://doi.org/10.1353/ncr.2004.0015

Zurr, Ionat, and Oron Catts. 2023. *Tissues, Cultures, Art.* Palgrave BioArt Series. Springer Nature.

CHAPTER 5

Techno-Rant of the Cyborg Elder

Faith Wilding

It is February 2024. Snow is falling. I am at my desk looking out on a wintry landscape of bare trees looming over a dark pine forest. Nature runs on light, air, and water, the most amazing computer of all. My friends and co-inspirators, Irina Aristarkhova and Hyla Willis, just left after we finished our two-day meeting in Northampton near the "independent living" apartment, which is my home for the foreseeable future. In an encouraging gesture, they bought me a new computer that looks much like the simple one on which I first wrote in the early '70s. I have not written on a computer for some years now, yet my fingers seem to find the keys without my looking at them. I'm awed by that fingertip memory, which appears to be effortlessly connected to my thinking and remembering abilities. I'm equally amazed and delighted that my amaryllis bulb is going through a second blooming, flaunting two plump buds slowly opening in the pale February sun. It is a miraculous example of "cellular intelligence," contained in a brown bulb that urges plants into growth and blossoming in rhythm with the lengthening sunlight.

Yesterday, we met at my friends' hotel for an initial conversation about our research interests, and to discuss some of the issues we want to cover in this book, including: aging and technology; the difficulties of "keeping up" with the always rapidly changing technologies of AI (artificial intelligence); my habit of kicking balls down the road, and tipping electric

H. Willis et al., *Night Sweats: Cyberfeminist Practices*, Palgrave BioArt,
https://doi.org/10.1007/978-3-032-06955-9_5

scooters over on Providence sidewalks just so I can hear them "complain." But first we went shopping to buy a computer for me! Oh dear goddess! Now I have to deal with a computer again! For decades, I wrote articles, letters, manifestos, and angry rants on my first simple Mac, which was much like a typewriter. Then I gave away my computer and began writing an autobiography in cursive on note-paper. Now, it seems, I will be writing again on a machine—and about a machine—that has scarily advanced since that first computer, and whose amazing memory and AI will spring into action the minute I touch a key (grrrrrrr), and will remember everything I've typed on it without me even asking it to do so (grrrrrrr! Hey, I thought I'm the boss here!).

I am thinking back to all the ways and means of written and recorded knowledge, and all the thinking and making I've done since I was a child, extremely fond of books, who learned to read at age 4. I would sit outside the bathroom door with my book, while my mother relaxed in the tub, and ask: "Mummy, what does 'F-L-O-W-E-R' say? What does 'P-L-A-N-T' say?" etc. And my dear Mum—a gifted schoolteacher—would sound it out for me. She taught me that one could make up all kinds of words out of a few letters strung together, and I began to understand how language works—that it is pieced together of sounds denoted by differently shaped letter-forms that represent meanings. And, equally amazing, I found that simple drawings of round, geometric, or plant-like forms and lines also held meanings that could be made into descriptive words and generate feelings, memories, and ideas. Thus, for me, drawing and writing were closely linked, and to this day, I am grateful that I was taught cursive writing early on in first grade. Years later, the Illuminated Manuscript became my favorite book form, and I've seen and studied many stunning examples in museums in the US and Europe.

How does this way of learning to write and read compare with my experience of learning how to use, first a manual typewriter (non-electric), and then an electric typewriter, and then a computer keyboard that had all the letters of the alphabet on it, as well as signs and symbols galore? For sure, the computer is a magical, marvelous MACHINE, as we tap out our business letters, our family jokes and news, our sorrows, loves, wishes, desires, our fun and friendship, riches and poverties, and our endless anxiety that the magic machine has changed everything, and already may have stored way more power and knowledge than we have. But does it have "life," can it think beyond algorithmic rhythms? Can it write wondrous poetry? Has AI changed the definition of what we think of as the knowledge, communication, and feelings, which cursive handwriting represented? Is it still true that my computer is "just a useful machine that will do what I tell it to" as it was in the beginning? Or have the computer, and Artificial Intelligence—and all that is connected with it—begun to take over what used to be the work of human hands, brains, hearts, and feelings? Many of us are wary of machines that seem to be smarter than we are, that create too much change by using technologies that can completely undo long-established ways of making, writing, living, and being. It seems "natural" that we should be wary of our computers, while also becoming deeply dependent and attached to them, often feeling closer to the machine than to our friends and families. Given that the computers of today can hold far more data, memories, information, and "knowledge" than even the most brainy human, why shouldn't we be jealous of them, scared of them, and/or in awe of them, and hold them above many human powers? Computers have become part of us physically, mentally, and personally, but so far, they are not sentient. When I hear people saying "I love my computer," "I couldn't live without my computer," or "my computer knows so much more than I do," I hesitate. I remember learning cursive writing in second grade, and being absolutely enthralled by making big loops on my "Ls and Gs" and coloring them in so they looked special. I practiced over and over until I was satisfied with my handwriting. I was amazed that my Mum could look at all these loops and strokes and "read" them, make sense of them. Much later, as I was able to study illuminated manuscripts in museums, I began to understand what a vast step in human life, and knowledge of all kinds, "writing" and "drawing" (writing is a kind of drawing as it is mark-making) represented.

I'm trying to approach my computer as a "memory machine," a machine that links me to my own memories as well as to vast histories of writing, communication, knowledge distribution, languages, and all that these entail. At the same time, I'm mourning what we have lost by using it for daily communications that used to be made by the human hand and voice: a recipe, a love letter, condolences, a school essay, a poem. Like many children, I was drawing long before I started writing. I scribbled, colored, drew blobby forms that stood for flowers, trees, or people—and then precise representations of leaf-forms which became my favorite subject (and still is). Nowadays, I'm thrilled to see my grand-nieces and -nephews still doing that kind of drawing and fundamental mark-making, as well as trying to tap out "stories" on their kiddie computers, and hopefully learning to write cursive also. Handwriting, like drawing, is personal; it comes from the body; it is intimate, and represents a deep history of language learning, communication, passing on family histories, experiences, and knowledge. Can technologies of machine writing—and our writing on machines—achieve similar feelings? Can we easily adopt whole new ways of connecting bodies, minds, and hearts using machines? Personally, I know how important these communication machines are to me, living as I do now, physically far from family, friends, and the "everyday world" of cities, travel, etc. However, I draw every day, and I wrote the first draft of this chapter by hand, in ink on lined paper.

In her fierce book *Glitch Feminism*, Legacy Russell (2020) lays out new ways of thinking about the relationship between bodies and machine technologies. She is not looking for an intimate and just relationship between body, gender, and technology. Rather, she suggests, there is no longer a divide between the body and technology. Interestingly, she describes the "shape-shifting" we can do with/through our machines: "...on line you can be anything you want to be," she writes; and "The digital machine world provides the potential possibility of creating new worlds" (Russell 2020, 2, 57, 106). She also makes the important suggestion that the Internet is a space for marginalized voices, who find others like themselves there. Importantly, Russell notes that AI is already an "existential threat"—for example, to the movie industry—with new "photogeneration software" that creates visual effects no longer made in the field with cameras. While this is a threat to movie industry workers, it allows independent creators to "create content they could not before." Thus, Russell argues that the digital machine-world does not discriminate or classify its users but allows them full creative freedom of mind and imagination. I have been thinking about this a lot, but am still not fully convinced that this is a "new ability" that transcends what the human imagination and perception can do without machines.

I am currently living in a community of mostly women, spanning in age from their 70s to their 90s. Many of them own computers that they use primarily to stay in contact with family and friends, do their taxes and business communication, and, importantly, to stay in touch with the "news," play games, and watch movies. Often at breakfast, someone will start a conversation by telling about what she's seen on her computer that morning, and others will immediately chime in with bits of information that they've picked up. Almost nobody seems to read the actual *New York Times* newspaper, which also arrives every morning. Instead, they read it online, because it is tedious to read the close, fine print, and not as

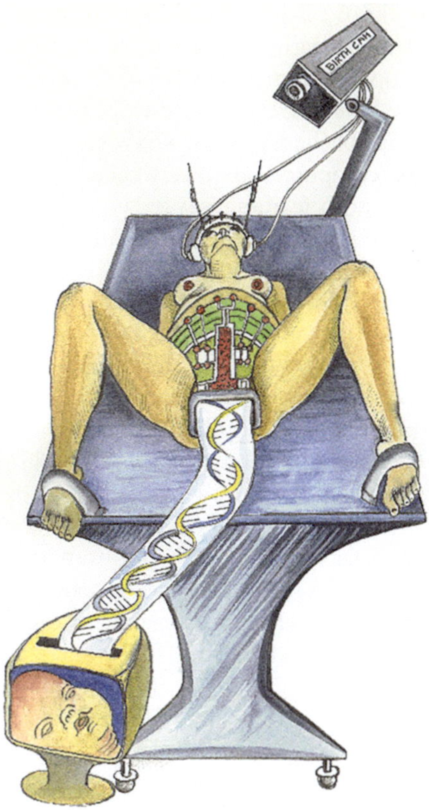

Fig. 5.1 A fictional Birth Cam and fetal DNA monitor drawn by Faith Wilding for *SmartMom*, a Net Art project that satirically "detourns" military technology developed for remote battlefield medicine to the uses of pregnancy surveillance (subRosa 1998).

"personal" as their computers are. I prefer to read the actual newspaper—an object that I'm very familiar with, and that I can carry around with me, and read almost anywhere. Perhaps it is also because I've always been addicted to written objects, be they books, magazines, or letters. Whether handwritten or printed, I want to hold them in my hands, and be able to turn the pages back and forth at will, and "feel" the writing and page formats, as familiar, indispensable visual, readable objects. Needless to say,

I feel the same way about reading actual printed books that I can hold in my hand, become familiar with, and take to bed with me.

Why are some people (like myself) wary of their computers—or rather, fear what the rapid, wholesale, even worldwide, adoption of the computer means to our long-established ways of knowledge attainment and communication? Why did computers so quickly become the universal "magical machine" that people both fear and desire, and which is now required in workplaces all over the world, and found in most homes and businesses? The ubiquity of the computer in homes, schools, offices, and workplaces points to a world that is more and more dependent on "instant" local and international communication and AI for its businesses, power politics, banking, and healthcare. The Internet and computers were not initially intended only for personal communication, knowledge-sharing, collaboration, etc. Like the radio and telephone communication networks that suddenly became widespread during the Second World War, the interplanetary networks of communication now being created by leading countries are far from being used mostly for purely scientific knowledge. Similarly, "social media" is often far from being "community building" or "social." Lately, young people are turning away from "social media" as it often builds no real sense of community or sociality, but rather leads to online shopping, misinformation, and bullying. Withdrawing from the "attention economy" (i.e., "texting" etc.) means engaging more closely and actively with your own community and environment.

These days, I've been drawing a lot and experiencing a renewed awareness of mark-making as an ancient and rich tool of communication and expression. Today, we still hear news stories of the discovery of charcoal or red mud marks and drawings made on the walls of caves thousands of years ago. The desire for mark-making is a deeply human urge that is both for communication and for recording presence and existence: "I am here, I existed." I'm reading a lot too, mostly novels that I have read before, and magazine articles about new technologies, and local and world politics. While drawing, I listen to NPR (National Public Radio) for news, as well as its many longer programs covering local issues and worldwide news and affairs. I have not watched the TV in my apartment since moving here, but we watch "independent films" on DVDs every night from one of the residents' personal collections. I watch with a handful of community members, feeling a little sad that most of the others are in bed watching their TVs alone. I remember that in Paraguay in the 1950s, we listened to music and concerts that had been recorded on big, thick gramophone records. We used a sharp thorn from a palm tree when the metal needle broke, and we had to crank up the record player by hand. Nowadays, a two-hour movie comes on a thin little disc you stick into a book-sized machine. Again, I ask: does the medium change the meaning or the quality of what is heard, seen, and expressed?

At mealtimes in the retirement home, there has been a lot of conversation about the upcoming [November 2024] presidential election, politics, and voting. It seems to me that if you have to pay taxes, you should be able to vote for the person who is going to be running the country and determining how to use my tax money, no matter where I was born. We often talk about AI and the pros and cons of its uses. Mainly, we worry about some of the damage the use of AI is already doing to artists, businesses, politics, etc. This is a primal worry many of us think about: "What do we do when our machines become smarter than we are?" "Can they be taught to make better art?" And what rules and laws, etc., can we all agree on that can be put in place to protect us all from the powers of the machines? Artists and writers have a lot to fear. The recent example of the use of so-called "generative AI" to "finish" a Keith Haring painting after his death from AIDS is a sad warning (Haring had left it "unfinished" to denote that the AIDS crisis was not "finished," write Elaine Velie and Rhea Nayyar 2024).

Artists are now being used to train Midjourney, a popular AI image generator: more than 16,000 "visual creators" are on the list, making robotics, animatronics, fake videos, etc. Some artists are using "generative AI" to create "dreamy body images"; or they join "Free the Nipple," a very "white women's" movement, that gives artists the "opportunity to create art free from conventional judgment" (Elkin 2023). This is very worrying, not least because it gives feminist art a troubling name, and once again makes the white female breast an elite object of commercial sexualization.

This brings me to questions of female aging, and technologies of immortality and death. What lives do we want and deserve? In my essay "Rant of the Menopausal Cyborg" (Wilding 2002/2003), I muse on aging and new technologies, machine immortality and embodied death: "What happens to an elder cyborg who can't use machines anymore? And whose body and life are deemed obsolete? Having a contingent cyborg body, along with your flesh and blood body, might be one way to go?"

And then there's the specter of AI. In a recent article in *The New Yorker* (March 18, 2024), Andrew Marantz discusses some of the problems and

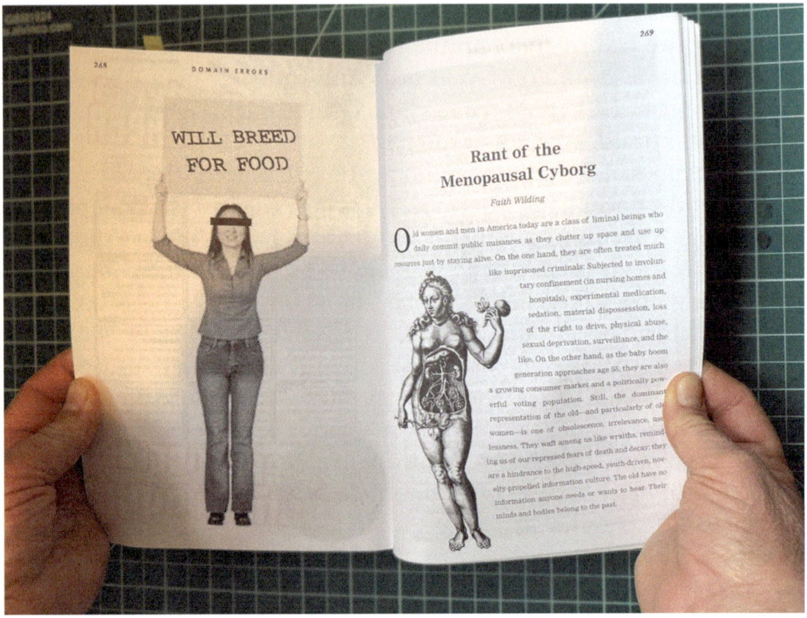

Fig. 5.2 A spread from *Domain Errors! Cyberfeminist Practices*, showing the opening page for Faith Wilding's "Rant of the Menopausal Cyborg." (Wilding 2002).

issues of AI, interviewing several people directly involved in the building and shaping of it. He pays close attention to the question of "the safety of AI," especially the age-old fear that humans might be able to make a machine that is more intelligent than we are, that might take over the control of humans and the world. He quotes a recent tweet from social media: "We must steal the fire of intelligence from the gods, and use it to propel humanity towards the stars," and notes that there is a group of so-called "techno-optimists" who oppose techno-pessimism and believe that the only way to go is forward to a utopian future fueled by AI (Marantz 2024, 42).

AI is a scary specter looming ever larger in the minds of many, most of whom don't have the slightest understanding of the machines that produce it. Many of those who do understand are deeply worried about "safety" and, as a survey of AI researchers revealed, "about half of AI researchers believed that the tools they were building might cause civilization-wide destruction" (Marantz 2024, 46).

Living in a retirement community is a strange echo of my early life history. My life began in a makeshift village in the forests of central Paraguay, where I was born into, and grew up in a religious commune—the "Bruderhof"—founded in Germany by a group of young people, many of whom had been part of the Wandervoegel (wandering birds) movement (see more on this in *Faith Wilding's Fearful Symmetries* edited by Shannon Stratton in 2018). The commune, which was committedly pacifist, fled Hitler's Germany during WWII and initially settled in England. There, my conscientious objector British father found them and decided to join them, bringing along with him my mother, who had just graduated from Teachers College, and his own mother and younger brother and sister.

England did not welcome Germans during the war, and eventually, the entire community had to find a country that would allow them to settle and follow their own pacifist faith and community way of life. This turned out to be Paraguay, a small, land-locked country in the middle of South America, where they had to build a functioning village in the midst of deep forests and endless grassland "campos." Fortunately, some of the commune members were trained engineers, farmers, woodworkers, etc., as well as highly educated Europeans. I was born there in 1943, about two years into the commune's settling in Paraguay, and just after its first steam engine had arrived. It was set up to provide ten hours of electricity a day, to power crucial woodworking and house-building equipment and serve a tropical disease hospital and the makeshift, straw-roofed, mud-walled, white-washed communal housing. The power was cut at ten o'clock at night and started up again at 6 a.m. Most of the steam-powered electricity was used to keep the turnery and woodworking machines going, as that was the commune's only way to make all the furniture needed for housing, wagonbuilding, and producing turned goods. Eventually, each home had one central lightbulb hanging from the ceiling in the middle of the living room. If one wanted to read in bed after lights-out, one had to finagle a flashlight and batteries somehow. As a child, I remember being intrigued by batteries and their power—they just seemed like magic, and of course, in the history of producing electric energy and much more, they still are used all over the world, especially in countries that do not have central electrical systems, and certainly no AI.

It is the summer of 2025, and my co-inspirators Irina and Hyla are visiting again during my birthday. Once again, my ability to use the computer to work on this essay has been hindered by unwanted and uninvited software updates that seem impossible to bypass. We read through a printed copy together out loud, and I am left with this question: WHERE is AI if I need it now? It is by now "everywhere," yet it is nowhere to be found to serve my real needs with these updates and the multitude of other tasks that computers condition the human body and mind to perform. The cellular memory of my amaryllis and the feeling of my hands on a keyboard are tangible forms of memory retrieval—intelligent and connected to the body. AI has neither the magic of the flashlight battery cells from my childhood nor the intelligence of my flowering amaryllis.

References

Elkin, Lauren. 2023. *Art Monsters: Unruly Bodies in Feminist Art.* Penguin Random House.

Marantz, Andrew. 2024. "Among the A.I. Doomsayers." *The New Yorker,* March 11. https://www.newyorker.com/magazine/2024/03/18/among-the-ai-doomsayers

Russell, Legacy. 2020. *Glitch Feminism: A Manifesto.* Verso.

Stratton, Shannon R., ed. 2018. *Faith Wilding's Fearful Symmetries.* Intellect Books.

Velie, Elaine, and Rhea Nayyar. 2024. "Ethical Questions Arise After AI 'Completes' Keith Haring Painting." *Hyperallergic,* January 8. https://hyperallergic.com/865291/ethical-questions-arise-after-ai-completes-keith-haring-paintin/

Wilding, Faith. 2002. "Rant of the Menopausal Cyborg." In *Domain Errors! Cyberfeminist Practices.* Edited by María Fernández, Faith Wilding, and Michelle M. Wright. Autonomedia. http://refugia.net/domainerrors/DE3i_rant.pdf

CHAPTER 6

subRosa Manifestations: A Chronology of Particular Projects, with Notes

Hyla Willis

This chapter aims to provide the reader with an overview of subRosa's projects, and to share some details that are not typically included in a curriculum vitae. Due to space and time considerations, this will not be an exhaustive accounting of the collective's output; however, I hope the information shared lends interesting insights into our enduring and complex collaborative practice.

On occasion, we had the privilege of being invited to write about our own work, or to glean insights about our work through the eyes of others, including numerous doctoral scholars and art historians. In the individual chapter references throughout this book, readers will find mention of numerous articles in which we took a deeper dive into certain of our projects.

Whenever possible, collective members sought to learn about the site, context, and situation—in the development phase—in which a new work would be presented, and to weave local topics and concerns into its overarching themes. subRosa projects are dense, complex, and sometimes share elements such as texts, props, and costumes with one another. Yes, we are an archivist's nightmare!

A commissioned work may have been revised and presented in a different setting with a slightly reworked title, such as with the *Sex and Gender Show*. Titles were sometimes recycled with slight changes to spelling or

H. Willis et al., *Night Sweats: Cyberfeminist Practices*, Palgrave BioArt,
https://doi.org/10.1007/978-3-032-06955-9_6

capitalization, effectively becoming wastebasket taxons to borrow a term from the sciences. Pamphlets and objects created as part of a larger commissioned work might have their own titles. For example, the *Cultures of Eugenics* booklet was created for *U-Gen-A-Chix* and reused frequently with *Cell Track. Constructa/vulva* was created for the *Knowing Bodies* installation but was later shown on its own. *Cell Track* was created as a standalone wall piece, but was also used as a backdrop for the *Epidermic! DIY Cell Lab.* The *Sex and Gender Ed Workbook* was a frequent prop in other performances or used as a giveaway at performative lectures. The *Knowing Bodies* title of our first exhibition installation was repurposed for a workshop in Winnipeg that resulted in a wonderful limited edition *ProActiva* poster design. Camel case was frequently used as a humorous nod to web server file naming conventions, but we had no official proofreader, and inconsistencies can be found across many years of documentation. The works below are presented roughly in the order of their first presentation, under their most enduring title.

Commissioned works were often supplemented by subRosa members giving public lectures or workshops during the site visit(s). These supplemental activities were often experimental and interactive in nature, or resulted in materials collaboratively produced with participants. This was also an important stream of income for our projects, since we would use lecture fees to stretch always-thin arts funding. Some of these lectures and workshops might be considered to stand on their own as works of art and are included here, with other appearances left for curious and diligent historians to unearth if they wish.

I have tried to include details that, while they may seem random to the general reader, may provide interesting insights to fellow artists, activists, historians, and collaborators.

SmartMom (1998)

A parody of a corporate website that is dedicated to remote pregnancy monitoring. It advertises a "SmartMom Sensate Pregnancy Dress," a détournement of the Smart T-shirt concept being developed for remote battlefield medicine at an American research university during the period between America's Gulf Wars. *SmartMom takes the "guess" out of "gestation."* (see Fig. 1.4 in Chap. 1 and Fig. 5.1 in Chap. 5). The site was first launched late in 1998 and was redesigned again in 2000 and 2009 (http://smartmom.cyberfeminism.net/).

Keywords: assisted reproductive technologies, bio art, cyberfeminism, feminism, net art, tactical media, wearables

CeLLular mUTaTIONS (1999)

subRosa members launch their first website for the group and hand out small red biohazard bags, announcing the formation of the collective. The bags contain a printed manifesto and a pair of latex gloves (Fig. 6.1). The first distribution was at the *Next Five Minutes* tactical media festival (Amsterdam 1999). The action was repeated at *Performative Sites: Intersecting Art, Technology, and the Body*, a symposium at Penn State University (State College 2000).

Keywords: cyberfeminism, feminism, manifestos, net art, performance art, publications, tactical media

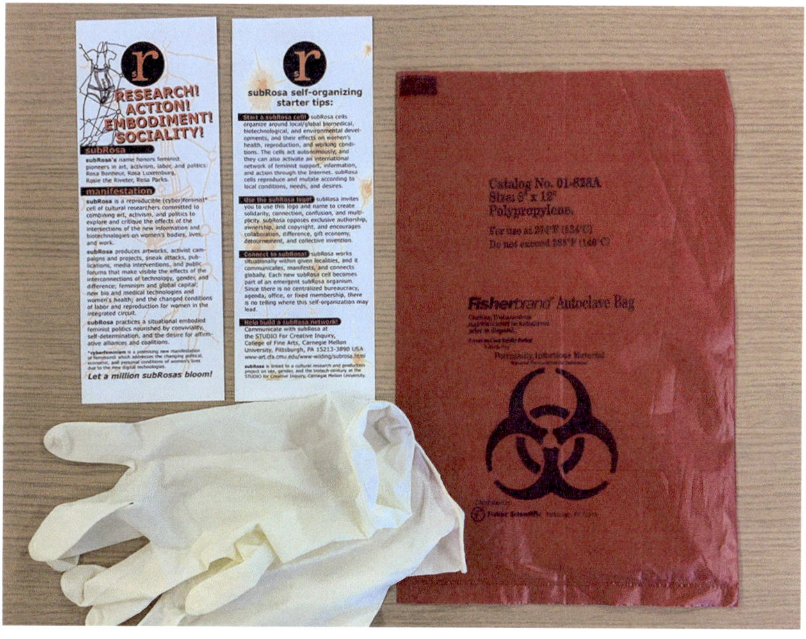

Fig. 6.1 subRosa's founding manifesto was distributed in red biohazard bags along with latex gloves to announce the formation of the group and the launching of its first website (subRosa 1999).

@*SecondOpinion* Newsletters (1999)

Collective members collaborate on the research, writing, and design of a provocative and humorous newsletter seeking to educate collegiate women being targeted by egg donor recruitment ads about the procedures, risks, and marketing language and compensation practices of the fertility industry. A second volume was produced four months later. The newsletters were distributed—without permission—in waiting rooms at the local women's hospital, on university campuses, and at a *Race for the Cure* event in Pittsburgh. The second volume featured a photograph of blindfolded women straddling a giant nest on the steps of the neoclassical building that was home to CMU's Biology Supercomputing Center. Included here is an outtake by photographer Beth Monoian, with Emily DeAraujo, Anna Skarbek, and Carrie Schneider seated (sans blindfolds in this outtake, and with Lucia Sommer included in the final blindfolded image). The nest and giant egg are later displayed in the *Knowing Bodies* exhibition described below (Pittsburgh 2000) (Fig. 6.2). Both volumes were offset printed in Pittsburgh in an edition of 1000. Volume 1 was six letter-sized pages printed in two colors. Volume 2 was a single color, quad folded tabloid broadside (subRosa 1999).

Keywords: assisted reproductive technologies, bio art, cyberfeminism, eugenics, feminism, installation art, performance art, publications, sex and gender

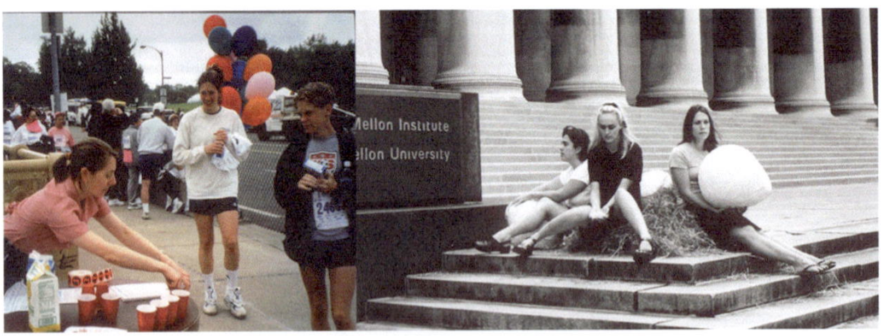

Fig. 6.2 A subRosa member hands out the first volume of the *@SecondOpinion* newsletter at a *Komen Race for the Cure* event in Pittsburgh (L). subRosa members conduct a photo shoot on the steps of a neoclassical building for use in the second volume of the newsletter (R) (subRosa 1999).

Vulva De/ReConstructa (2000)

A humorous video on the subject of aesthetic surgery of the vulva premiered at the *Die Verletzte Diva* exhibition at Taxispalais in Innsbruck, Austria. The infomercial-like video examines the phenomenon of "aesthetic surgery" of the vulva and, through parody, raises questions about the rhetoric of "choice" in body modification. A fictional patient meets with a plastic surgeon (Fig. 6.3) expressing her wish for an oversized, asymmetrical, "floppy" labia (subRosa 2000, 9:19).

Keywords: cyberfeminism, feminism, sex and gender, video art

Knowing Bodies (2000)

For the inaugural exhibition of the Regina Gouger Miller Gallery at Carnegie Mellon, featuring projects hosted by the STUDIO for Creative Inquiry (Pittsburgh 2000), subRosa created an installed environment featuring its web site, the oversized nest and egg constructed for the @ *SecondOpinion* newsletter photo shoot, a coloring table with select pages from the *Sex and Gender Education* workbook, a "gene pool" with petri-dish-encased assemblages floating in a kiddie pool, and a gallery wall

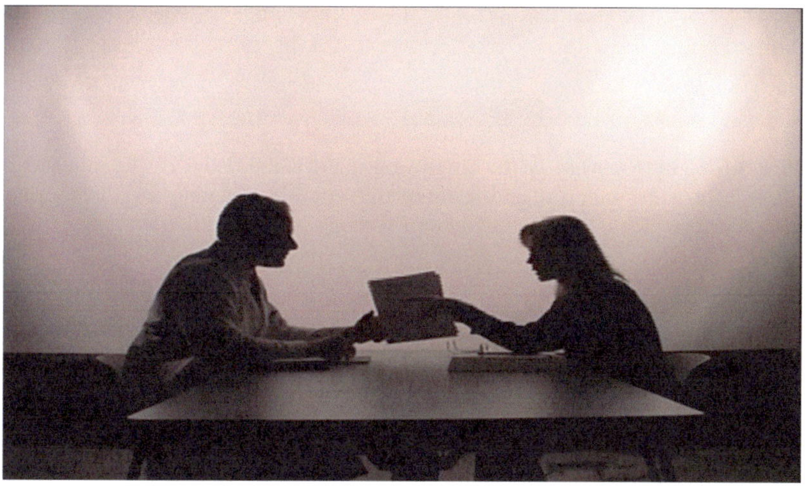

Fig. 6.3 "Interview scene," *Vulva De/ReConstructa*. A fictional labiaplasty patient explains her unorthodox requests to a surgeon (subRosa 2000).

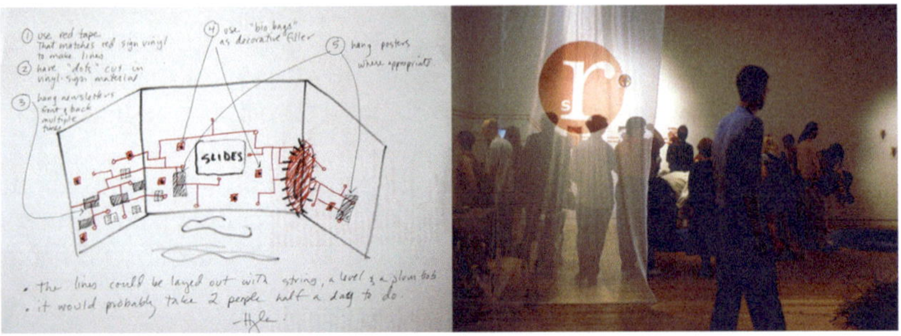

Fig. 6.4 Preparatory sketches and an installation view from the *Knowing Bodies* installation in *Fusion! Artists in a Research Setting*. The nest and oversized egg created for the *@SecondOpinion* newsletter can be seen in the lower right corner (subRosa 2000).

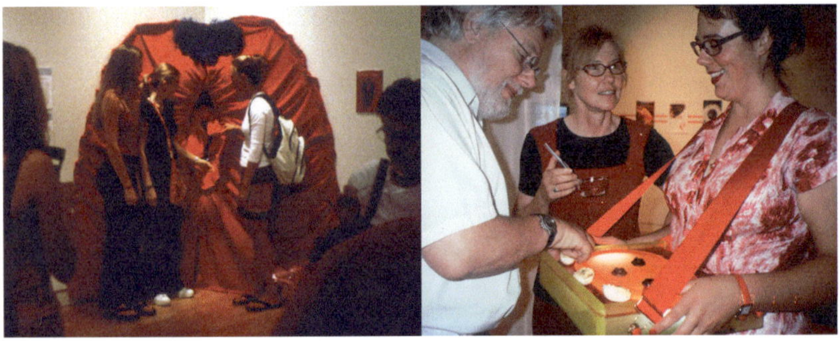

Fig. 6.5 The *Knowing Bodies* installation marked the debut of the interactive *Constructa/vulva* sculpture and featured cigarette girls roaming and distributing custom-made "Megabite Tasties" gummy candies (subRosa 2000).

featuring broadsides and videos created by the collective (Fig. 6.4). A central feature is the premiere of *Constructa/vulva*, a giant soft sculpture described below. During the opening event, "cigarette girls" with bespoke trays circulated the floor, handing out gummy candies molded to resemble genitalia and DNA strands (Fig. 6.5).

Keywords: assisted reproductive technologies, bio art, cyberfeminism, eugenics, feminism, feminist history of science, installation art, manifestos, net art, performance art, publications, sex and gender, video art

Constructa/vulva (2000)

An interactive super-sized sculptural performance is created as an element of subRosa's *Knowing Bodies* installation in the *Fusion! Artists in a Research Setting* exhibition [see Fig. 6.5] at the Regina Gouger Miller Gallery (Pittsburgh 2000). *Constructa/vulva* (subRosa 2000) is styled with a blue Merkin—a sly reference to the larger-than-life sexless *Stuffee* anatomical doll at Pittsburgh Children's Museum that helped inspire the project. Like *Stuffee*, *Constructa/vulva* comes with a wildly colorful collection of anatomical 'parts', in this case including many sizes and shapes of labia, cervix, urethra, and clitoris. Performers (in recent iterations, costumed with shiny arm pieces, to resemble specula) encourage and assist audience members in creating their ideal female genitalia by affixing their chosen parts to Velcroed surfaces. An instant photograph of the audience member with their creation is presented on the spot. Christina Hung led the design, construction, and draping of *Constructa/vulva* (Fig. 6.6).

Constructa/vulva was later shown in *Everybody!: Visual resistance in feminist health movements, 1969–2009*, curated by Bonnie Fortune. I-Space (Chicago 2009) and Carleton College Gallery (Northfield, MN 2011).

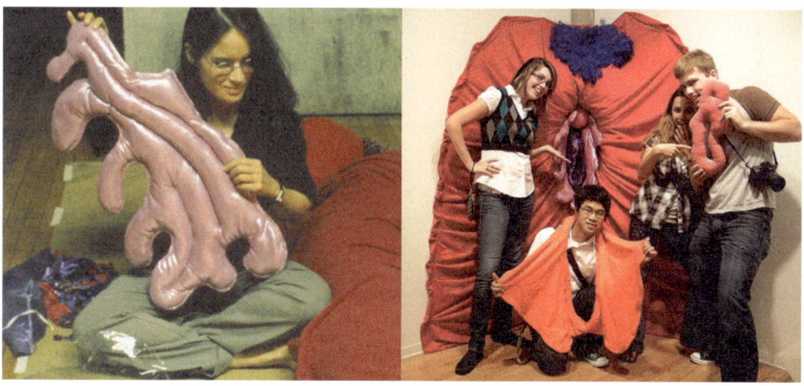

Fig. 6.6 *Constructa/vulva* took on a life of its own after the *Fusion!* exhibition in Pittsburgh. Hand-sewn anatomical parts were crafted from colorful fabrics, and users could interact with each other and the Velcro-attached pieces to design their own oversized genitalia (subRosa 2000).

Keywords: bio art, eugenics, feminism, feminist history of science, installation art, performance art, publications, sex and gender, social practice, wearables

The Sex and Gender Education Show (2000, 2002)

Styled as a secondary school sex education class peppered with product placements (a commentary on the commercialization of mathematics textbooks in the US), subRosa's first public performance employed time-honored, low-tech teaching methods to inspire critical thinking and knowledge sharing about the new Assisted Reproductive Technologies that were coming into wide use at the time. subRosa members posed as corporate and government agency representatives in a classroom takeover, registering class participants, collecting saliva samples, and assigning them to one of four groups named for the protein bases of DNA (T, A, G, & C), or a fifth, "dud" group (See Figs. 4.11 and 4.12 in Chap. 4). Following a sing-along to the tune of the "America the Beautiful," audience members were given a workbook (See Fig. 4.9 and 4.10 in Chap. 4) and an illustrated crash course consisting of mini-lectures on reproductive genetics and ART methods (Fig. 6.7).

In a hilarious final "repro-tech mixer," participants mingled to a cocktail hour soundtrack while seeking reproductive partner(s) matching their preferences to negotiate "making a perfect baby the biotech way." This included a rebellious breakaway group of Luddites desiring to have babies the old-fashioned, sexy way.

subRosa members spent a considerable amount of time scripting and animating the mini-lectures and creating logos to represent the fictional enterprises infiltrating the classroom setting. In doing research for the show, collective members looked at ARTs and their effects on female sexuality, reproductive choice, eugenics, and gender in the biotech century.

Sex and Gender in the Biotech Century was commissioned work for the *Digital Secrets: New Collaborations in Visual Art and Technology* conference at Arizona State University (Tempe, AZ 2000).

The Sex and Gender Education Show was a slightly revised and renamed version created for the *Hardware, Software, Wetware & Women* conference at St. Mary's College of Maryland (St. Mary's City 2002).

Keywords: assisted reproductive technologies, bio art, cyberfeminism, eugenics, feminism, feminist history of science, net art, performance art, publications, sex and gender, tactical media, racism

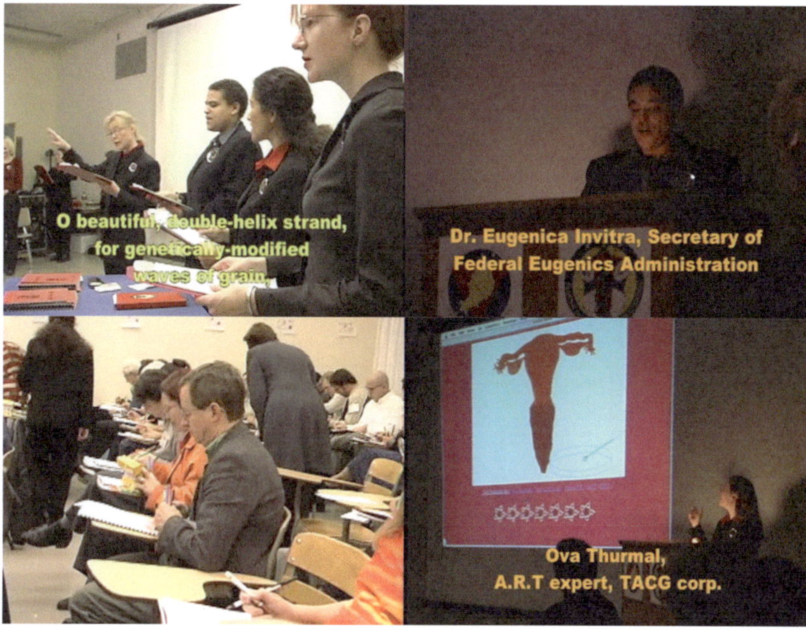

Fig. 6.7 Stills from video documentation of *Sex and Gender in the Biotech Century* at *Digital Secrets*, depicting a sing-along, mini lectures, and participants completing workbooks to prepare for a genetic mixer (subRosa 2000).

Expo EmmaGenics (2001)

An immersive and interactive performance disguised as a trade show centered on the growing market for Assisted Reproductive Technologies in the US. Visitors could browse tables that featured fictional products, mix and mingle with each other, and experience occasional disruptions for medical demonstrations and raffles. At the venue entrance, visitors had to cross a staged picket line with protestors shouting and carrying placards demanding "Autonomy for the Mother!". Inside, a media- and advertising-dense environment used humor to draw comparisons and contrast between EU and US access, regulations, marketing and attitudes towards IVF and other ARTs (Fig. 6.8). Although subRosa was unfamiliar with the term *kayfabe* (from US professional wrestling) at the time, the performance relied on a willing suspension of disbelief on the part of its

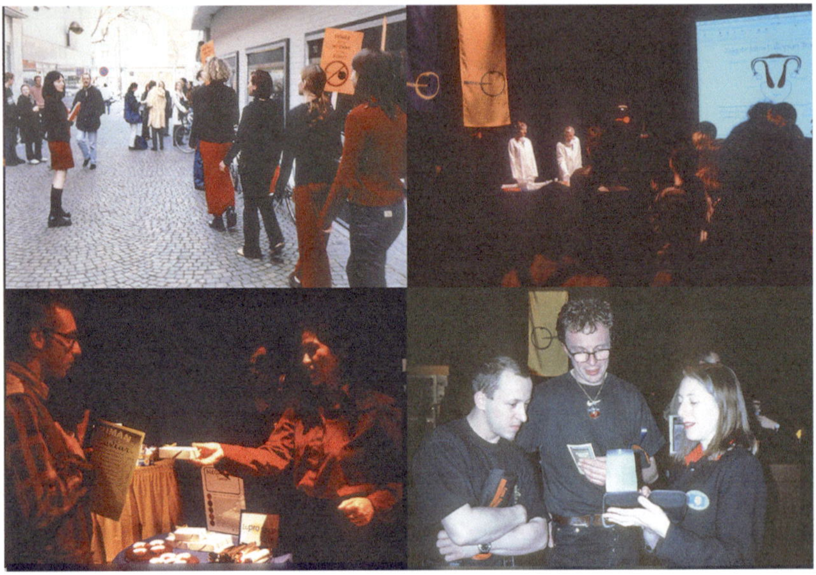

Fig. 6.8 Scenes from *Expo EmmaGenics* depicting, clockwise from upper left, a staged protest, a lecture and demonstration about zygote collection for ARTs, a demonstration of the "PalmXY Matefinder" app, and samples of "Megabyte Tasties," "Lupro Bars," and "Human Caviar" in packaging reminiscent of Starbucks branding (subRosa 2001).

audience as subRosa performed the roles of fertility doctor, product vendor, or "volunteer from the audience" for demonstrations explaining how eggs and sperm are harvested for ART procedures. Speculative/fictional products included human caviar, baby-cams trained on cells dividing in the IVF lab, and a dating-type app on a PalmPilot to help identify potential biological or social co-parents.

> *The genetic future of our planet is too precious to be left in the hands of nature and chance! Western Europeans have been slower than Americans to embrace the promises of new bio-, genetic-, and reproductive technologies. Expo EmmaGenics is designed to show Europeans how to take advantage of the efficiency, usefulness, genetic enhancement, and choice that these technologies offer* (From the *Intermediale* festival program booklet).

Expo EmmaGenics was created for *Art Happens! Intermediale*, a festival of "electronic theatre and interdisciplinary theatre, media performances and performance installations" and performed at TiC Studiobühne (Mainz, Germany 2001). *Intermediale* was integrated into the *7th International Performance Studies Conference* (PSi7) scheduled concurrently at the University of Mainz.

Keywords: assisted reproductive technologies, bio art, cyberfeminism, domestic spaces, eugenics, feminism, installation art, net art, performance art, publications, sex and gender, tactical media

Knowing Bodies residency & *ProActiva* poster (2001)

subRosa contributed to a week-long collaborative research residency in Winnipeg that brought together international women working in the arts, humanities, health care, and the sciences to discuss the conditions of our daily lives, work environments, and the health-care and health insurance systems we each navigated. Among many activities, workshop participants collaborated with subRosa to produce a limited edition poster (Fig. 6.9) mocking the pharmaceutical advertisements commonly seen in American doctors' offices and waiting rooms at the time. The "Witch Root" plant is a collage of several different historic plant and fungus drawings and has appeared in a few other subRosa publications and as a wall mural in the *Propositions and Undoings* installation (subRosa 2011).

The *Knowing Bodies* residency was held at St. Norbert's Art and Cultural Center (Winnipeg, Canada 2001). The poster was later displayed in *Pittsburgh's Avant Garde*, Irma Freeman Center for the Imagination (Pittsburgh 2024).

Keywords: bio art, cyberfeminism, feminism, feminist history of science, manifestos, publications, social practice, tactical media

Becoming Autonomous Zones (BAZ) scarves (2001)

In response to a call for posters for the *Very Cyberfeminist International* International (Hamburg, Germany 2001), subRosa hand-decorated a series of Tyvek squares as multi-purpose *BAZ* scarves: useful for fashion, shelter, storage, and manifesto distribution (Fig. 6.10). The title *BAZ* was a riff on the popular Autonomedia book *Temporary Autonomous Zones* by Hakim Bey. subRosa's *Becoming Autonomous Zones* was introduced in the "Refugia Manifesto for Becoming Autonomous Zones [BAZ]" printed in the *Domain Errors!* (subRosa 2002, 277) book. The scarves were displayed alongside

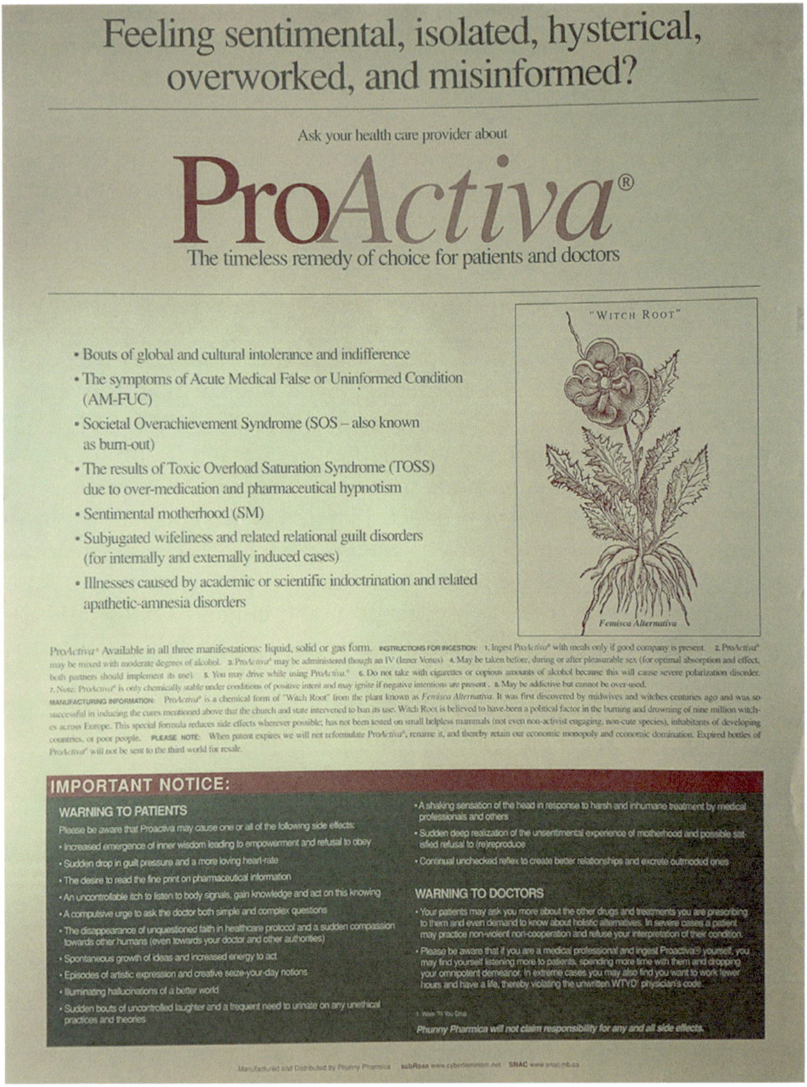

Fig. 6.9 A satirical pharmaceutical advertisement produced by participants in the *Knowing Bodies* residency. 2-color offset on green paper (subRosa 2001).

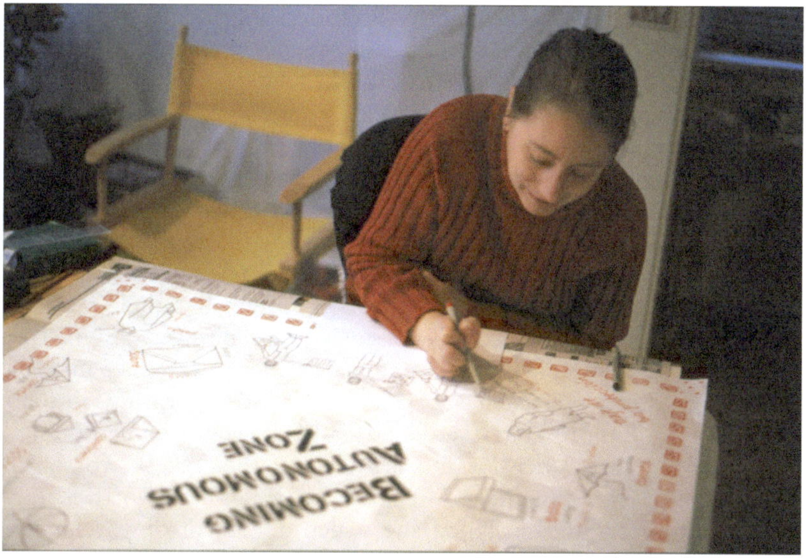

Fig. 6.10 A subRosa member adding hand-drawn details to a Tyvek *Becoming Autonomous Zones (BAZ)* scarf (subRosa 2001).

a lecture and presentation about the material Tyvek and its parent company, DuPont, holders of the first patented living organism: OncoMouse. The scarves were also included in *Cyberfem. Feminisms on the Electronic Landscape* at Espai d'art contemporary de Castelló (EACC) (Spain 2006).

Keywords: cyberfeminism, domestic spaces, feminism, feminist history of science, intellectual property, manifestos, social practice, tactical media, wearables

U-Gen-A-Chix (originally *US Grade AAA Premium Eggs*) (2002)

Originally staged inside the student center of a large public university, *U-Gen-A-Chix* engaged students directly in critical conversations about eugenics as it relates to contemporary genetic engineering of humans, animals, and plants. Tandem performance booths were set up in a high-traffic area: one dispensing information on human egg donation and Assisted Reproductive Technologies, and the other offering taste tests of chicken-flavored biscuits made with genetically modified ingredients. After taste-testing the biscuit, participants gave live video interviews (staged as a mock focus group) about their willingness to eat genetically engineered foods if they enhanced energy performance during exams, and offered their opinions about the widespread use of GMOs and related biological and social eugenic tendencies. The earliest version of this performance premiered at Bowling Green State University (Ohio 2002), as *Domain Errors!* was wending its way through Autonomedia's publishing process. Originally titled *US Grade AAA Premium Eggs*—it was advertised with a pink flyer designed by Steffi Domike that depicted a young female tennis player, deep in a volley shot of a flume of chicken eggs (Fig. 6.11). Agents wearing lab coats posed as egg donor recruiters and held a tabling event where interested visitors could surf a website and learn about their "flesh worth" in the emerging fertility market. After completing a "flesh worth form," participants were given a signed and sealed document on parchment, designed to resemble a stock certificate.

On the occasion of Ryan Griffis' *YouGenics 2.0: Exploring the Social Implications of Genetic Technologies* exhibition at Southwest Missouri State University (Springfield 2003), the show was renamed *U-Gen-A-Chix*, revised, and performed several more times in public spaces on campuses where it would be common to encounter marketing tables, though more typically for internet service providers, military recruitment, or student clubs.

Revisions to *US Grade AAA Premium Eggs* included adding the second table to distribute fresh-baked snacks (Fig. 6.12) and a media team to conduct mock focus group interviews. College campuses were then common recruiting grounds for egg donors, and compensation was often correlated with a semester's tuition at the school [i.e., egg donors at community colleges were routinely compensated far less than donors at top research universities.] Selling one's eggs remains illegal; however, fertility banks may compensate donors for their time. The interviews provided an opportunity for reflective conversation with visitors, and the tapes would be played at some later iterations of the performance (Fig. 6.13), but the interviews were not gathered for any research project they were a performance: part of the corporate ruse.

The expanded performance also marked the introduction of the *Cultures of Eugenics* pamphlet, which contained parallel texts comparing the role of

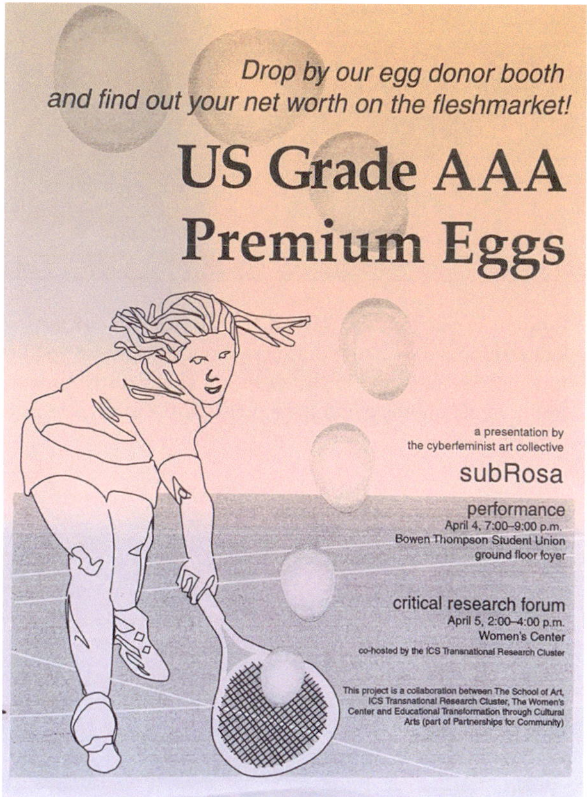

Fig. 6.11 The letter-sized flyer advertising *US Grade AAA Premium Eggs* performance and other campus activities that were part of subRosa's visit.

females in the avian and human fertility industries, and a timeline and texts introducing a history of eugenics. The pamphlet's cover featured art from an outdated Stanford-Binet test: one that was used to determine human intelligence, and a contributing factor in the forced sterilization of some individuals at the height of the American Eugenics movement (See Fig. 2.2 in Chap. 2).

The original web-based "Flesh Worth Calculator" became a broadside designed to mimic a common tax form (See Fig. 2.3 in Chap. 2), with versions in English and Spanish that were subsequently used in the

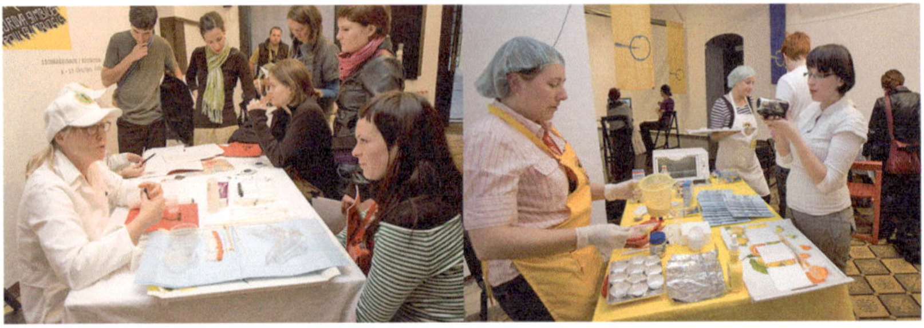

Fig. 6.12 Tandem tables that are part of the *U-Gen-A-Chix* performance. On the left, an agent attempts to recruit young egg donors and help visitors determine their worth on the flesh market. On the right, an agent distributes freshly-baked treats containing genetically-modified ingredients and *Cultures of Eugenics* booklets (subRosa 2008).

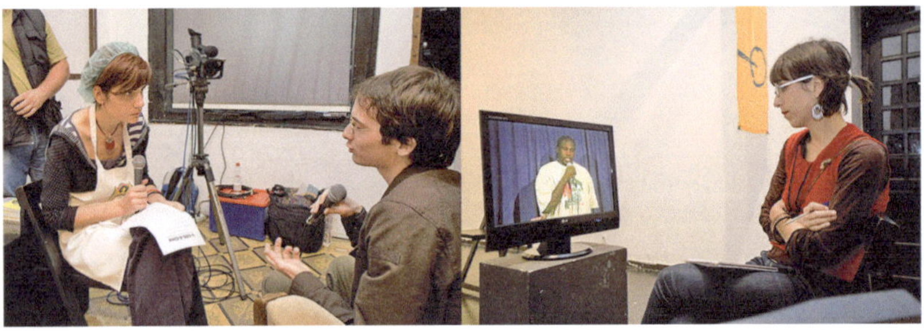

Fig. 6.13 Visitors to the two *U-Gen-A-Chix* tables are invited for a "focus group" style interview about the products and information they have just encountered. Interviews from prior performances are displayed on a monitor in a different part of the gallery (subRosa 2008).

International Markets of Flesh performances (subRosa 2003), along with ornate "Flesh Worth Certificates."

US Grade AAA Premium Eggs was renamed *U-Gen-A-Chix* to emphasize links to both eugenics and to the universities where female students were being recruited as egg donors and where *U-Gen-A-Chix* was typically presented. At Syracuse University (New York 2005), media arts students assisted with the performance, taking up the role of the "crew" conducting mock focus group interviews.

The show was presented at Galerija Škuc in Lubljana as part of *Animals_Nature_Culture, 14th International Festival of Contemporary Arts—City of Women* festival (Slovenia 2008), and it was remarkable to note that participants' access to and attitudes toward ARTs (highly regulated and available as part of socialized health care) and GMO foods (all but banned and deeply unpopular) were inverted in a mirror-like fashion from US participants' access and attitudes (ARTs highly unregulated and then seldom covered by insurance in the US, while GMO ingredients were already uncontroversially widespread in popular processed food brands). Four young artists from Zagreb—whom subRosa had met months earlier during their *Look! Listen! A Week With|Out Women* residency (subRosa 2008) were guest performers who helped run the tables and film focus group interviews. They included the Wo Kolektiv and Iva Kovač, who years later would become program director for City of Women. The differences in site, situation, and audience background at each performance were woven into our ad hoc dialog and addressed in broadsides and microsites that were created as performance "props" that extended the reach of our research and creative efforts beyond the life of the performance itself, which typically lasted for 4–6 hours.

Keywords: agriculture, assisted reproductive technologies, bio art, cyberfeminism, eugenics, feminism, performance art, publications, sex and gender, tactical media

Biopower Unlimited! (2002)

Based on the success of *US Grade AAA Premium Eggs*, subRosa was invited to create a new project for the Bowling Green State University 23rd Annual *New Media & Art Festival* (Ohio 2002). The festival happened to coincide with the annual tech fair on campus, a sprawling tradeshow-like event whose primary purpose was to sell computers and promote technology clubs to freshmen. This led us to stage a campus intervention covering both events. *Biopower Unlimited!* Included a custom *Cultures of Technology* map—developed in consultation with students, faculty, and community activists—that revealed the intersections of biological, agricultural, and digital technology cultures on the BGSU campus and surrounding rural farm town. On site at the tech fair, the subRosa team staffed a booth where participants could complete an online "Biopower Profiler" comparing how they allocated their labor power and leisure time for a

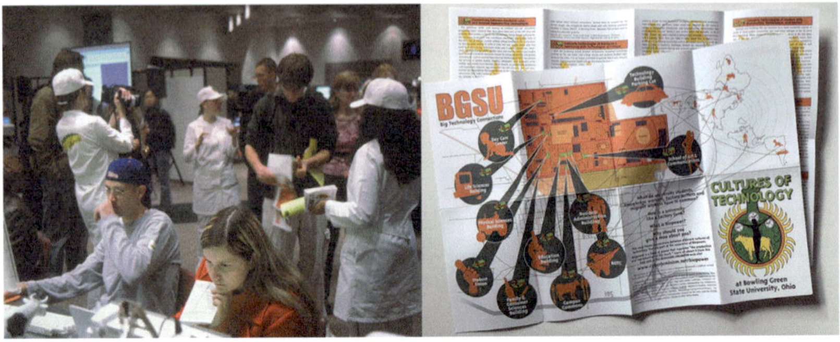

Fig. 6.14 subRosa members dressed in costume greet visitors to their booth in the BGSU annual technology fair. Students could complete an online "biopower profiler" and pick up a *Cultures of Technology* map distributed at the fair and in mailboxes across campus (subRosa 2002).

total biopower calculation. "Consultants" helped participants analyze their results and gave advice, such as reframing their athletic scholarships as labor, not leisure (Fig. 6.14).

The maps were distributed at the technology fair and via interactive mailboxes strategically placed across campus (Fig. 6.14).

One topic included on the map was the proliferation of high-tech dairy "Factory Farms" in the area, which the team had documented easily, simply by showing up in uniform with video cameras and clipboards, but without prior permission. The map raised issues that were further personalized by the biopower profiler at the trade show. Anyone opening a mailbox to retrieve a map was greeted by a small motion-sensitive mooing cow inside (Fig. 6.15).

The lab coat costumes used in earlier performances got a refresh with the addition of a BIOPOWER logo across the back, matching ballcaps with an embroidered alchemical emblem, and ribbon-striped trousers (See Fig. 1.3 in Chapter 1 and Fig. 4.7 in Chapter 4).

Along with these creative interventions, subRosa conducted a graduate colloquium and video screening on the issues of biopower and technology during our campus visit.

Additional images from the performance were reproduced in *The Interventionists: Users' Manual for the Creative Disruption of Everyday Life* (Thompson and Sholette, eds, MIT 2004). A portion of the map is

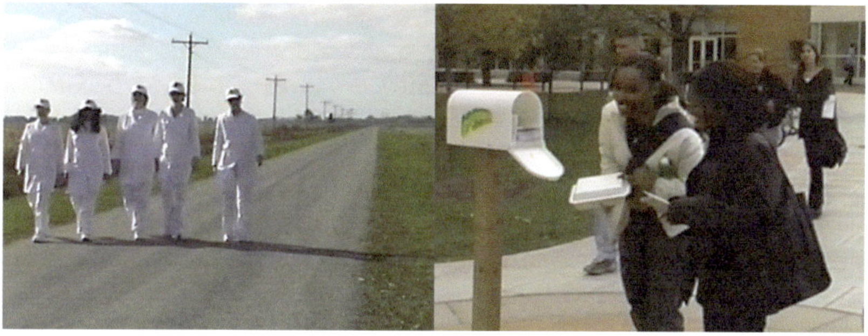

Fig. 6.15 Two video stills from documentation of the *Biopower Unlimited!* project at BGSU. On the left, subRosa members in costume walk down a road during a research trip where they dropped in on local dairy factory farms. On the right, two BGSU students open a white mailbox on campus to discover a mooing cow inside, one of several set up as distribution points for the *Cultures of Technology* map (subRosa 2002).

reproduced in *Experimental Geography: Radical Approaches to Landscape, Cartography, and Urbanism* (Thompson, Melville House 2009).

Keywords: agriculture, bio art, cyberfeminism, experimental cartography, feminism, performance art, publications, social practice, tactical media

Domain Errors! Cyberfeminist Practices (2002)

Please see Chapter 3, "Domain Errors! A Virtual Roundtable" (See Fig. 3.1, Chap. 3 and Fig. 5.2, Chap. 5) for a discussion of this subRosa book project.

Keywords: assisted reproductive technologies, bio art, cyberfeminism, domestic spaces, eugenics, feminism, manifestos, publications, sex and gender, racism

Matrixial Technologies Tactical Media Lab (2003)

A collaboration between subRosa members, Irina Aristarkhova, Margaret Tan, Adeline Kueh, and affiliates of the Cyberarts and Cyberculture Research Initiative at the National University of Singapore (NUS) and

Lasalle College of Art in Singapore. At the time, stem cell research was being curtailed in the US and many other countries while Singapore—already a highly technologized country—was in a race to become a pre-eminent global biotech/bioinformatics supplier and life sciences research center.

Collaboration participants sought to map and make visible local and global flows of stem cell lines and analyze how bioinformatics was producing new collective social, intellectual, and political bodies. We looked for signs of how old and new relations of colonization, especially endo-colonization at the genetic level, are being reproduced in the economy of stem cell informatics.

Activities of the two-week Tactical Media Lab included a visit to an endangered species cloning lab, a seminar at a private umbilical cord stem cell bank, interviews with chief scientific researchers and lab managers, a bio-science methodology seminar with university scholars, guest lectures, a tour of the NICU unit at the KK Women and Children's hospital where cultural differences in birthing practice are tended to, and a two-day creative workshop—"Cloning Cultures"—organized by subRosa for art students and faculty (See Fig. 1.1 in Chap. 1). subRosa also met with the Women in the Arts group AWARE.

Keywords: agriculture, assisted reproductive technologies, bio art, cyberfeminism, domestic spaces, eugenics, feminism, feminist history of science, intellectual property, performance art, sex and gender, social practice, trafficking, racism, political love

International Markets of Flesh (IMF) (2003, 2005)

An audience-participatory performance and collective mapping of the global trafficking in human organs and tissues, *International Markets of Flesh [IMF]* first took place on the former altar of a seventeenth-century convent church (Fig. 6.16). The title was an intentional riff on the International Monetary Fund, a humorous detail that may have been lost on the non-English-speaking primary audience for the show. Through participatory activities, informative lectures, and the manipulation of life-size organ sculptures and diagrams, the audience learned about the growing international demand for transplantable organs and tissues, and the political, social, and medical consequences these demands have created. Participants wrote personal stories and recorded popular rumors about

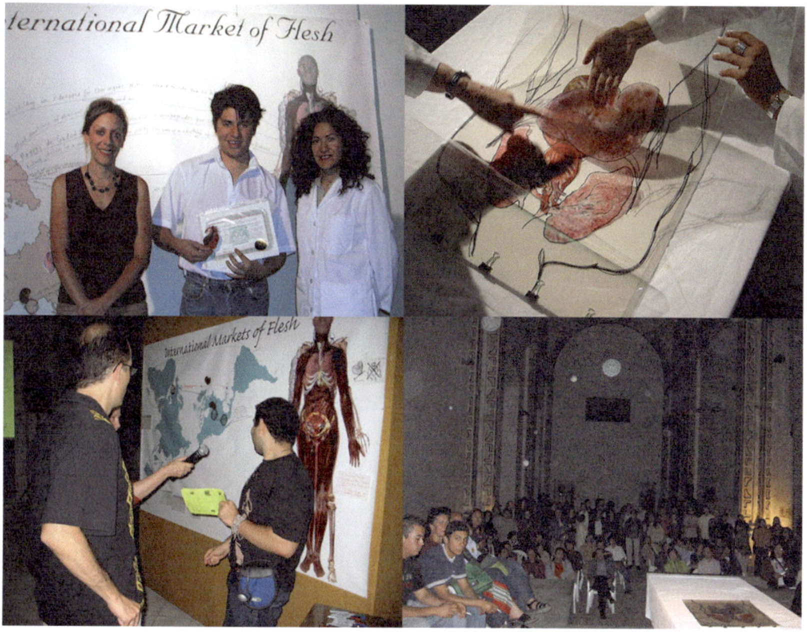

Fig. 6.16 Four still images from *International Markets of Flesh*, depicting—clockwise from the top left—a translator and a subRosa performer flanking an audience member who is holding up his "Certificate of Flesh Worth;" a layered anatomical drawing of life-sized human organs drawn with ink on mylar; audience members assembled around the church altar where the performance was held; and two audience members sharing stories out loud as they add details to a large collaborative map (subRosa 2003, 2005).

organ harvesting and trade on a large Dymaxion-style world map. The visual accumulation of facts, fiction, testimony, and stickers with colorful drawings of commonly traded organs effectively demonstrated the dominant flows of the flesh market worldwide, with demand coming from the North/West and supply coming from the South/East. Performers and audience members also discussed changing ideas about the value of human life in the age of genetically engineered, globally distributed, and patented human body parts, and filled in a form estimating the net worth of their body parts and labor to receive a Certificate of Flesh Worth. In the second performance, to encourage audience interaction, we rewarded participants with colorful biopower currency for contributing ideas and stories to the map.

IMF was performed once in Ciudad de México and once in Mérida. On both occasions, texts and translations were provided in English and Spanish with the help of on-site translators and some subRosa members' fluency in both languages. Both in México and Mérida we became more aware that despite our efforts to be inclusive with our texts, many indigenous languages are still used daily in North America. For English speakers based in the US, it is common to forget that Spanish is also a colonial language! At the time Mexico was debating its laws about organ donation and seeking ways to combat trafficking. Unlike the US, in Mexico individuals must opt-out of organ donation upon death. The US requires people to opt-in. When discussing examples of kidney trafficking in Mexico and the US, our audience raised parallels with international adoption. While there are many examples of above-board international adoptions, several people in the audience cited reports of young women from poor villages whose babies disappeared after a heavily-drugged childbirth, and the women later gaslit about having been pregnant at all.

The performance and map explored the idea of parallel texts by overlapping and intermingling scientific knowledge with Internet rumors and personal lived experiences of the audience members. The mapping aspect was in part to tease apart where these realities overlapped and intersected on actual geographies, and where they might be notably absent.

IMF was first performed by Laleh Mehran, Faith Wilding, and Hyla Willis for *XI International Performance Art Festival: Out of Focus*, ExTeresa Arte Actual, (México 2003), then by Carolina Loyola-Garcia, Faith Wilding, and Hyla Willis for *Arte Nuevo InteractivA'05*, Patio Central del Centro Cultural Olimpio (Mérida 2005). Ephemera from the shows was included in *A Studio of Their Own: The Legacy of the Fresno Feminist Experiment, 1970*, Conley Art Gallery (Fresno 2009). *Frakcija Performing Arts Journal* (Constant Capture, no. 43/44 2007) contains an illustrated article by subRosa—"International Markets of Flesh: Mapping Flows of Human Organs and Tissues"—with a long description and documentation of this project.

Keywords: bio art, cyberfeminism, experimental cartography, feminism, intellectual property, performance art, publications, sex and gender, social practice, tactical media, trafficking, racism

Cell Track: Mapping the Appropriation of Life Materials (2004)

Cell Track: Mapping the Appropriation of Life Materials is a flexible-scale wall installation and accompanying web site that examines the privatization and patenting of human, animal, and plant genomes within the context of a history of eugenics. It has been exhibited alone and in combination with other subRosa projects, sometimes with site-specific content modifications.

Cell Track draws attention to the increasing separation between the bodies that produce stem cells—and other genetic material—and the medical and pharmaceutical products derived from them.

Maternal body cells and tissues—such as eggs, placentas, fetuses, and umbilical cord blood—have become valuable "raw materials" extracted and accumulated for emerging biotech industries. This has opened new pathways for corporate science to profit from the manipulation and control of life via the privatization of DNA sequences, engineered genes, stem cell lines, transgenic organisms, and the like. *Cell Track* gives a historic overview of the privatization of biology while also raising the possibility of activist, feminist-inspired experimental research and labs accessible to amateurs, artists, independent scientists, and non-profit researchers conducting research and sharing knowledge for the public good.

subRosa's most-exhibited work of art, the first version of *Cell Track* was developed for *Bio-Difference: The Political Ecology, Biennale of Electronic Arts Perth* (Australia 2004) and included the *Cell Track* website and a large wall graphic featuring a figure with ambiguous genitalia and a body occluded by raised red discs citing examples of successful and unsuccessful legal challenges to the privatization of bodily tissues (Fig. 6.17). Adjacent to the figure was a long stripe of reflective gold petri dishes protruding from the wall, above marketing slogans for biotech start-ups and below a large printed quotation from Vandana Shiva.

For Ryan Griffis' *YOUGenics 3.0* exhibition at the Betty Rymer Gallery (Chicago 2004), *Cell Track* was revised to include a large Dymaxion-style map projection behind the figure and a horizontal band of cards outlining a timeline of events marking biology's changing relationship to private property (generously designed by Stephanie Rothenberg). The installation furthermore served as a visual backdrop

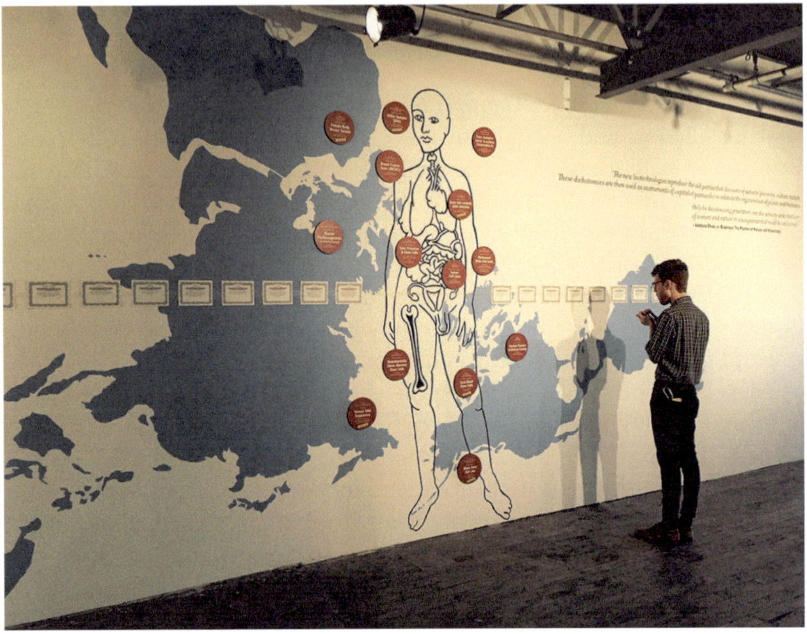

Fig. 6.17 An installation of *Cell Track* at the *Cellular Control* exhibition in Montreal, Canada (subRosa 2015).

to the *Epidermic! DIY Cell Lab* performance that was also included in the *YOUGenics* show. The "Cultures of Eugenics" pamphlet, originally developed for *U-Gen-A-Chix*, (subRosa 2003), is frequently distributed alongside *Cell Track*.

Cell Track has also been included in *Deliciously Disposable Earth*, Three Rivers Arts Festival Gallery (Pittsburgh 2008), *Soft Power. Art and Technologies in the Biopolitical Age*, Amarika Project (Vitoria-Gasteiz, Spain 2009), *BPLTC I: Cellular Control*, Eastern Bloc (Montréal 2015), and *subRosa Atlanta Anti-University* at Murmur Gallery (Atlanta 2016).

Keywords: agriculture, assisted reproductive technologies, bio art, cyberfeminism, eugenics, experimental cartography, feminism, feminist history of science, installation art, intellectual property, net art, sex and gender

Can You See Us Now? ¿Ya Nos Pueden Ver? (2004)

This two-part interactive installation for *The Interventionists: Art in the Social Sphere*, MASS MoCA (North Adams 2004) mapped the intersections of women's material and affective labor in cultures of production in North Adams, MA, and Ciudad Juárez, Mexico, and investigated the similarities and differences of economic, cultural and every-day life effects of the outsourcing of labor and globalization on these towns and on their local female labor force. Large, aerial wall maps of North Adams and Ciudad Juárez featured oversized map pins denoting "points of view," and flanked a "forensic floor" that concealed objects, texts, and clues beneath loose boards (Fig. 6.18). Visitors were encouraged to discover connections between the aerial maps, the contents beneath the floor, and a third printed "tourist map" distributed in the space (Fig. 6.19). Women engineers from nearby Smith College were invited to create a floor diorama explaining the material culture of capacitor manufacturing as part of a class project. Also displayed were five posters by contemporary Mexican artists, expressing concern and outrage about the continuing murder and disappearance of women in Ciudad Juárez, lent by the Center for the Study of Political Graphics. The "Clothing Tag Map" was displayed separately, in the foyer of the Museum, and encouraged visitors to cut the tags off their clothing and pin them to a Dymaxion-style world map, according to the location where each garment was manufactured (Fig. 6.20). Thus, visitors

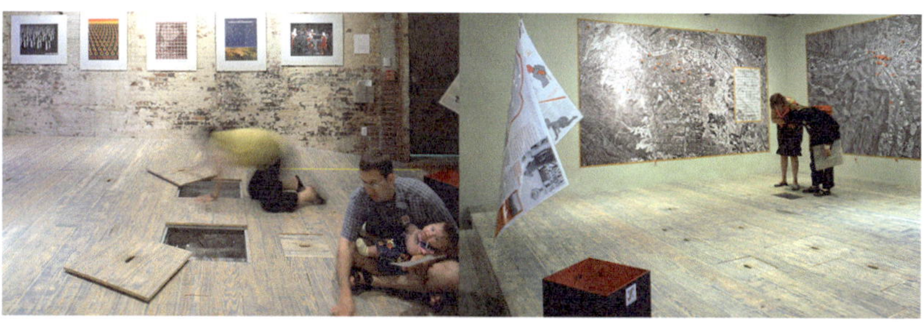

Fig. 6.18 *Can You See Us Now? ¿Ya Nos Pueden Ver?* featured a large "forensic floor" where visitors could discover themed dioramas beneath the floorboards. Reprints of Mexican posters calling attention to femicide in Ciudad Juárez were framed on one wall, while two other walls depicted large aerial maps of North Adams—where the exhibit was staged, in a former capacitor factory—and of Juárez—where the factory work moved to, making way for the museum.

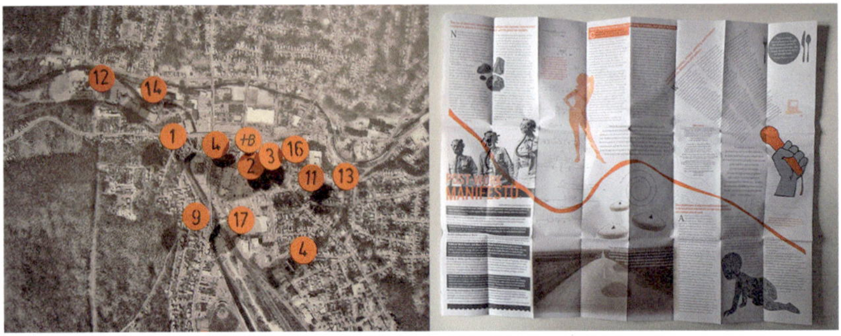

Fig. 6.19 Oversized map pins helped visitors draw connections between the items in the dioramas beneath the floorboards and the living and working conditions shared by women in the two locations featured in the exhibition. For example, the networks supporting solidarity between workers and shelter from domestic abuse and industrial pollution. A printed broadside that folded down to resemble a tourist road map offered further insight into subRosa's research for this project, linking the history of women's labor in North Adams and Ciudad Juárez (subRosa 2004).

actively explored and demonstrated their own participation and complicity in globalized labor conditions.

The "Clothing Tag Map" was shown again in *Thought Crimes: The Art of Subversion*, Diverse Works (Houston, TX 2005) and the many tags collected from both showings were donated to artist Wendy Osher, who refashioned them into a diaphanous tapestry, *Continental Drift* (Osher 2009).

Keywords: cyberfeminism, domestic spaces, eugenics, experimental cartography, feminism, feminist history of science, gentrification, installation art, social practice, tactical media, trafficking, racism, political love

Epidermic! DIY Cell Lab (2005)

In this performance-driven installation created for Ryan Griffis' *YOUGenics3* at the Betty Rymer Gallery (Chicago 2005), gallery visitors were able to view examples from a classic experiment in recombinant DNA bacteria; learn to streak petri dishes; practice basic vocabulary around adult and embryonic stem cell cloning; see yogurt-making as an example of kitchen-based cell culturing; and create "recombinant"

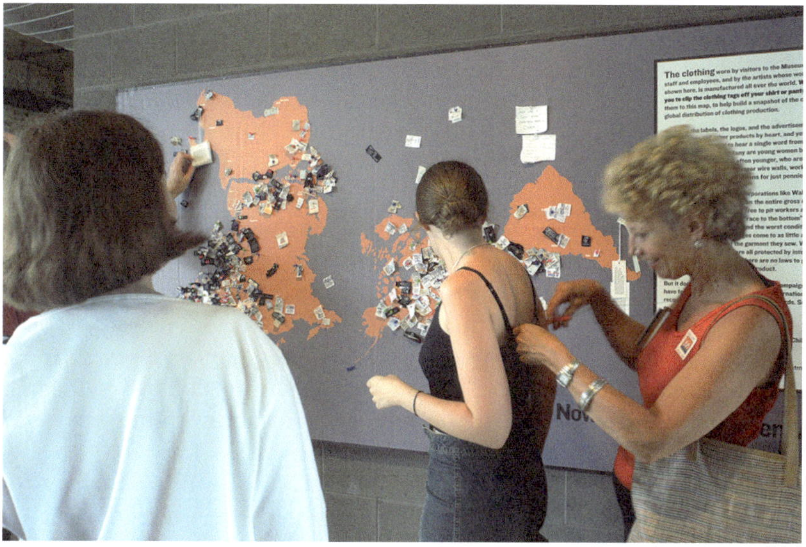

Fig. 6.20 A Dymaxion-style map was placed in the museum lobby, where visitors were invited to cut the tags off their clothing and pin them to the point of origin mentioned on the tag. As the tags accrued, they revealed the geographic clusters of clothing manufacturing around the world (subRosa 2004).

paper collages from historic and contemporary scientific and alchemical imagery (See Fig. 1.2 in Chap. 1). Thus, visitors were exposed to different ways in which science is represented to the public, and got a taste of public knowledge production in action! The project's goal was to familiarize visitors with the general processes and concepts of the recombinant stem cell biotechnology that underlies genetic engineering. *Cell Track* is a separate wall installation (seen in the background) that was used as a backdrop for this performance. Art students Elena Marcevska (nee Jovanova) and Liz Rosenfeld were guest performers and the show was revised for *Soft Power. Art and Technologies in the Biopolitical Age*, curated by Maria Ptqk at Amarika Project (Vitoria-Gasteiz, Spain 2009).

Keywords: assisted reproductive technologies, bio art, cyberfeminism, domestic spaces, eugenics, feminism, feminist history of science, performance art

Yes Species (2005)

Yes Species is a 20-minute multimedia tableau that imagines the meeting of three philosophers "performing gender differently" in a forest clearing: a live DJ mixing recorded vocalizations; a second performer standing in vats of red and green ink, blowing into vellum text balloons; and a third manipulating scrolls of text and dispersing copies of the *Yes Species* book with freshly-imprinted covers. First developed for *1-0-1 Intersex: The two-gendered system as a Human Rights Violation* at NGBK (Berlin 2005), the work explores both the performativity of gender as theorized by Judith Butler, and the hope that "things can be thought differently" as Luce Irigaray has suggested. *Yes Species* was a collaboration between subRosa and James Pei-Mun (now Wu) Tsang and is also a small book containing texts that were distributed as part of the performance, with translations in German and Spanish (See Figs. 4.1–4.3 in Chap. 4).

The performance was staged again for *Cyberfem. Feminisms on the Electronic Landscape*, Espai d'art contemporani de Castelló (EACC) (Spain 2006).

Keywords: cyberfeminism, domestic spaces, feminism, performance art, publications, sex and gender

Love Is Strong as Death (2006)

Love is Strong as Death was first staged as a feast for the five senses, inspired by the inscription (after the Song of Solomon) on the memorial Carrie Tower at Brown University. Participants congregated around festive tables of locally-sourced home-cooked food (Fig. 6.21) and were invited to "listen, imagine, and respond to aspects and acts of political love." Commissioned by Brown University's Art Department, this work hearkened back to Plato's Symposium by taking on the theme of "political love." The body politic is addressed through the experience of the human sensorium and the embodied situation of sharing food, wine, and thinking together (Fig. 6.22). The feast performance produced ephemera and objects that were then left in the venue as a sculptural installation.

In Transit: from Object to Site, David Winton Bell Gallery, Brown University (Providence, RI 2006). *Pathogeographies (or, other people's baggage)*, Gallery 400, (Chicago 2007). In documentary form in *Propagation*,

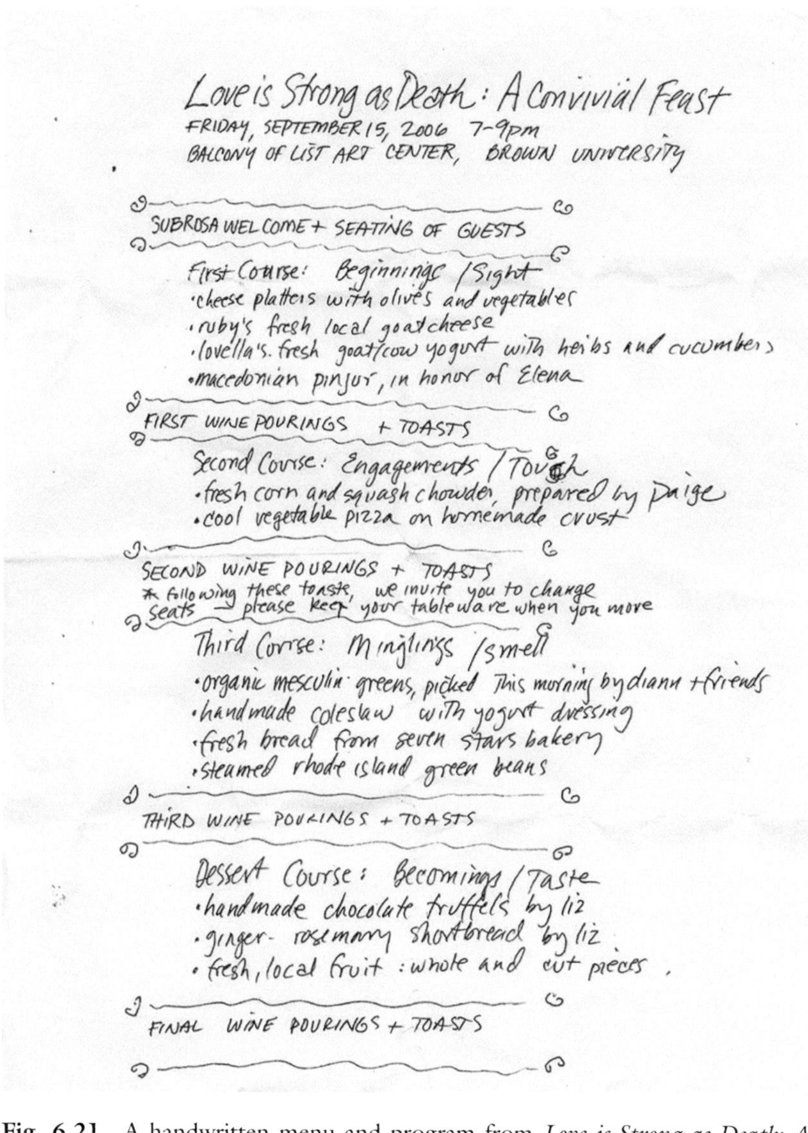

Love is Strong as Death: A Convivial Feast
FRIDAY, SEPTEMBER 15, 2006 7–9pm
BALCONY OF LIST ART CENTER, BROWN UNIVERSITY

SUBROSA WELCOME + SEATING OF GUESTS

First Course: Beginnings / Sight
·cheese platters with olives and vegetables
·ruby's fresh local goatcheese
·lovella's fresh goat/cow yogurt with herbs and cucumbers
·macedonian pinjur, in honor of Elena

FIRST WINE POURINGS + TOASTS

Second Course: Engagements / Touch
·fresh corn and squash chowder, prepared by Paige
·cool vegetable pizza on homemade crust

SECOND WINE POURINGS + TOASTS
* following these toasts, we invite you to change seats — please keep your tableware when you move

Third Course: Minglings / Smell
·organic mesculin greens, picked this morning by diann + friends
·handmade coleslaw with yogurt dressing
·fresh bread from seven stars bakery
·steamed rhode island green beans

THIRD WINE POURINGS + TOASTS

Dessert Course: Becomings / Taste
·handmade chocolate truffels by liz
·ginger-rosemary shortbread by liz
·fresh, local fruit: whole and cut pieces

FINAL WINE POURINGS + TOASTS

Fig. 6.21 A handwritten menu and program from *Love is Strong as Death: A Convivial Feast* staged on the balcony of the List Art Center at Brown University and featuring locally-sourced ingredients. Artists Paige Sarlin and Liz Rosenfeld joined subRosa in preparing the food (subRosa 2006).

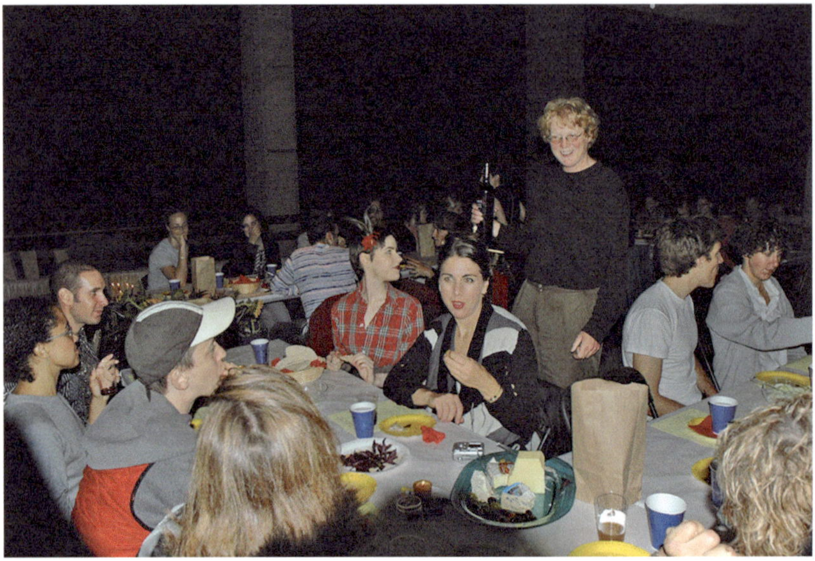

Fig. 6.22 Guests at the *Convivial Feast* had been invited to "listen, respond, and prepare a gift of discourse" throughout the dinner, in response to texts and ideas that subRosa had shared earlier, addressing ideas about the politics of friendship between women, and how "acts of political love" could be activated anew (subRosa 2006).

curated by Sabrina Raaf at Polvo (Chicago 2006) and in *Women Artists: Then and Now*, Track 16 Gallery, (Santa Monica, CA 2007).

Keywords: agriculture, domestic spaces, feminism, performance art, social practice, political love

Gestation: A Workshop and Map (2008)

A collaborative performance art workshop led by subRosa and Elena Marcevska at New York University Skopje (Macedonia 2008), with an interdisciplinary group of scholars, artists & NGO workers. Combining interdisciplinary art and theater skills, we created a tourist-like map to make visible the "hidden feminist histories" of Skopje, reframing some of the existing monuments and imagining possibilities for new ones. This

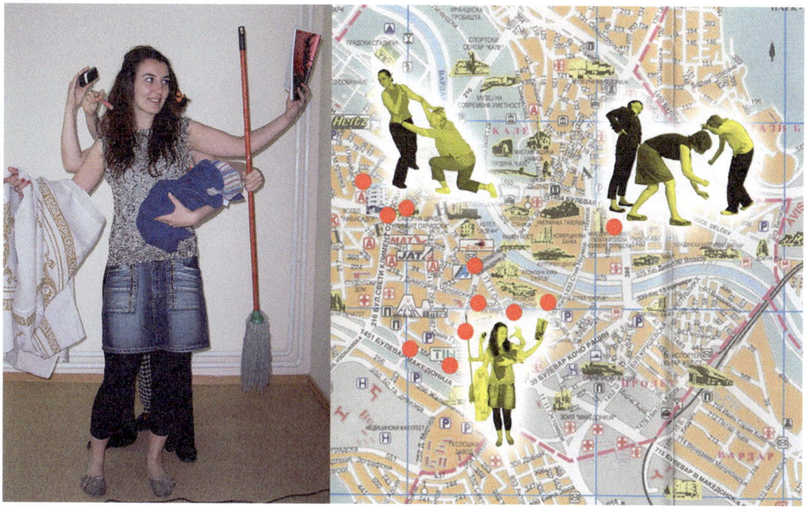

Fig. 6.23 Workshop participants explored the monuments of downtown Skopje together, then reimagined them as tributes to the hidden feminist histories of the region, combining storytelling, experimental theater, and visual design to overwrite a popular tourist map (subRosa and Marcevska 2008).

included historic and present-day concerns experienced by women of different ages and ethnic and religious backgrounds (Fig. 6.23).

Keywords: domestic spaces, experimental cartography, feminism, gentrification, sex and gender, social practice, tactical media, trafficking

Look! Listen! A Week With|Out Women (2008)

subRosa conducted a ten-day residency in Zagreb, where different groups of women were invited to daily gatherings to address a variety of concerns elicited through organized discussions about everyday life and political conditions. Evening salons were followed by a "manifestation" the next day in appropriate city locations (Fig. 6.24). Documentation of our activities was posted daily in WHW's Galerija Nova, along with notes about "What We Heard" and asking "What are Your Demands?"

The residency culminated in a performative feast at the Student Center Club, Teater &TD, University of Zagreb (Croatia 2008), during which

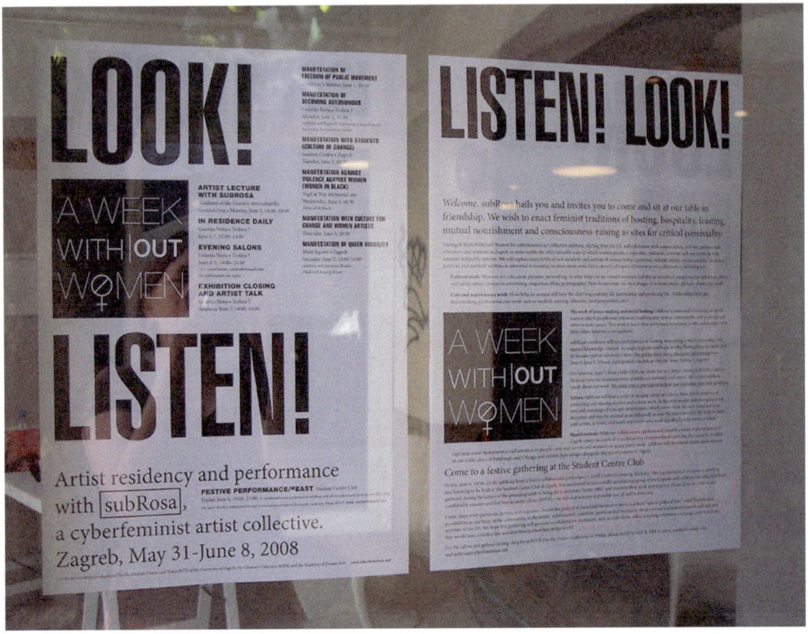

Fig. 6.24 The two-sided broadside advertising subRosa's *Look! Listen! A Week With|Out Women* residency and performances, seen on display in the storefront of the WHW Galerija Nova in Zagreb. The gallery served as a meeting point and display space throughout the week (subRosa 2008).

participants hailed one another, unveiled a plan for a "SubVersity," and revisited the intergenerational social and political connections that were made throughout the residency (Fig. 6.25).

Keywords: cyberfeminism, domestic spaces, feminism, gentrification, manifestos, sex and gender, social practice, tactical media, trafficking, political love

Out-Rage-Us-Joy (2008)

A poetic, performative installation, commissioned for an inter-generational feminist exhibition: *The Way That We Rhyme: Women, Art and Politics*, curated by Berin Golonu at the Yerba Buena Center for the Arts (San

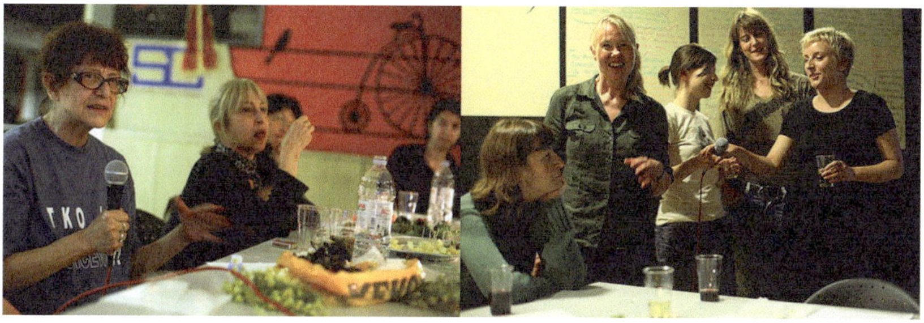

Fig. 6.25 *Look! Listen! A Week With|Out Women* ended with a performative feast where participants hailed one another and the connections that had been made during the residency. The intergenerational sharing of life conditions and political struggles between women who came of age on opposite sides of the Yugoslav wars was one of the most surprising and gratifying aspects of this project (subRosa 2008).

Francisco 2008) subRosa borrowed from the aesthetic forms and utopian gestures of the "Art Into Life" philosophy of Russian Constructivists, and the "Wall of Respect" painted in 1967 by Chicago-based black activists, to initiate discussions of feminist joy and solidarity. Performers rode on two 'rafts' made from painted canvas and discarded tires, billboard materials, and invited viewers aboard. Participants were invited to decorate paper "Cones of Respect"—shout-outs to inspirational figures that were pinned like barnacles on a "pier post" reminiscent of a Constructivist communications tower. The project demonstrated "the way that we work" in solidarity with each other and with other women, and invokes different social movements and philosophies that influence each of us (Fig. 6.26).

Keywords: cyberfeminism, feminism, installation art, performance art, political love

Miss|Placed Women (2009)

subRosa curated a "shift" of programming for the 15th gathering of *Performance Studies International* (PSi15) (Zagreb, Croatia 2009) entitled *Miss|Placed Women*, that included a panel discussion and performances (Fig. 6.27) featuring artists Elena Marcevska, Violeta Luna, Tanja Ostojic (with her similarly-titled performance *Misplaced Women*), Sanja

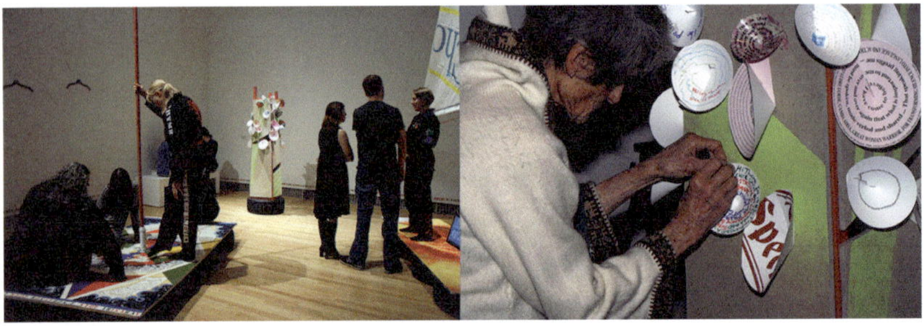

Fig. 6.26 For a show focusing on intergenerational feminisms, subRosa created two large rafts and a pier post floating on upcycled tires, depicting "the way that we rhyme," the curatorial theme inspired by Le Tigre song lyrics. During the opening, we greeted and conversed with visitors and invited them to contribute folded paper cones with shout-outs to the women who inspire them. Our hand-painted flight suits were a tribute to Russian Constructivist costume designs by Popova and others (subRosa 2008).

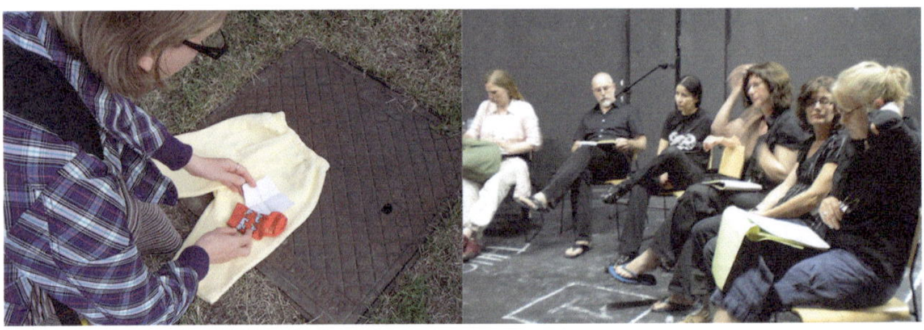

Fig. 6.27 Left: Elena Marcevska left infant's clothing and handwritten notes to map motherhood throughout downtown Zagreb. Right: performances and presentations from *Miss|Placed Women* were followed by a public conversation between the artists and audience, moderated by subRosa and guest Bettina Knaup.

Ivekovic, Iva Kovac, Roberto Varea, and the wo kolektiv. Our shift was in part a response to the 2006 PSi12 gathering in London, which carried the theme "Performing Rights," but that had failed to produce the timely letters needed for non-EU Europeans to garner the visas needed for travel to the conference. The shift was chronicled on a website [http://missplaced.refugia.net/] and in a special edition of *Frakcija* dedicated to PSi15.

Keywords: cyberfeminism, domestic spaces, feminism, gentrification, performance art, sex and gender, social practice, tactical media, trafficking, racism, political love

Bodies Unlimited! Becoming Feminist Scientists (2010)

Building upon workshops with Maria Ptqk and philosopher Paul B. Preciado, introducing the subject of biotech, subRosa led a workshop for artists exploring tactics and strategies for interdisciplinary knowledge production, research, creative experimentation, response, and representation. Participants shared visual presentations, kept shared lab books, and engaged with basic techniques of observation and transformation in the sciences (Fig. 6.28). Small groups then worked together to produce a final public presentation combining experimental art and science.

Created for *Soft Power II. Art and Technologies in the Biopolitical Age*, Sala Rekalde (Bilbao, Spain 2010).

Keywords: bio art, cyberfeminism, feminist history of science, social practice

Feminist Matter(s): Propositions and Undoings (2011)

Created for the *Pittsburgh Biennial* curated by Astra Suparak at the Regina Gouger Miller Gallery (Pittsburgh 2011), this project was intended to initiate collective inquiry about how feminist knowledge sharing might

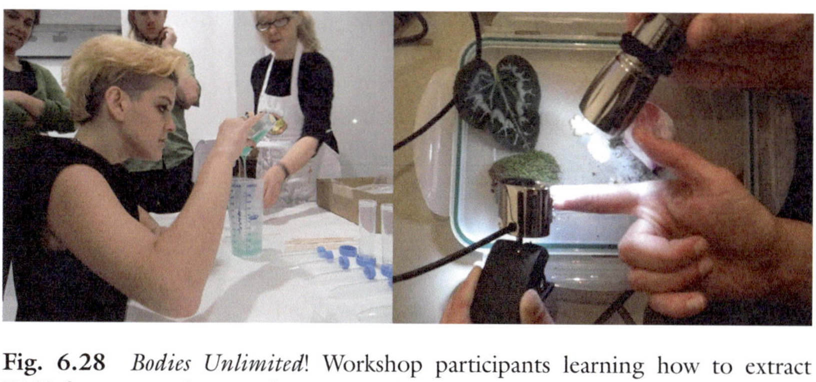

Fig. 6.28 *Bodies Unlimited*! Workshop participants learning how to extract DNA from a strawberry and reverse engineer webcams to become plant magnifiers (subRosa 2010).

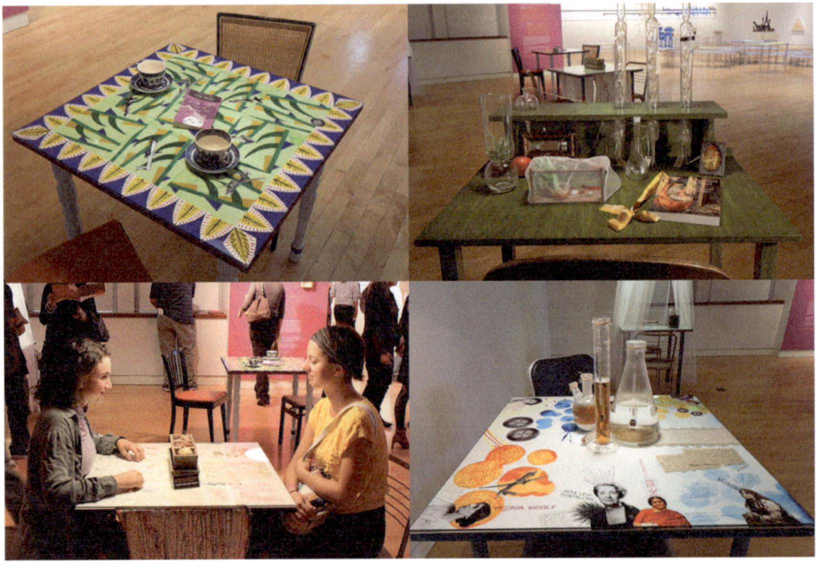

Fig. 6.29 Four of the tabletops participants could sit at in the *Feminist Matter(s): Propositions and Undoings* exhibition. Each two-person setting combined drawings, collage, and curated objects to reveal feminist histories of science and scientists (subRosa 2011).

inform "bench-side" approaches to scientific method and science pedagogy. Five small tables arranged in a cafe-like setting were inspired by Virginia Woolf's *Thoughts on Peace in an Air Raid*, in which she introduces the idea of "tea-table thinking" as a counter to boardroom or war room thinking (Fig. 6.29). The tables featured colorful and detailed drawings and collages under glass, with clusters of artifacts as centerpieces on top. A closer look revealed that each table was themed with important stories of women scientists, artists, and tinkerers who did groundbreaking work in their own kitchens or while relegated to subpar lab spaces, including Remedios Varo, Barbara McClintock, Vandana Shiva, Hypatia, Rita Levi-Montalcini, and others.

Gallery visitors were invited to sit at the different tables, alone or with another person, to think and talk about the themes, histories, and ideas

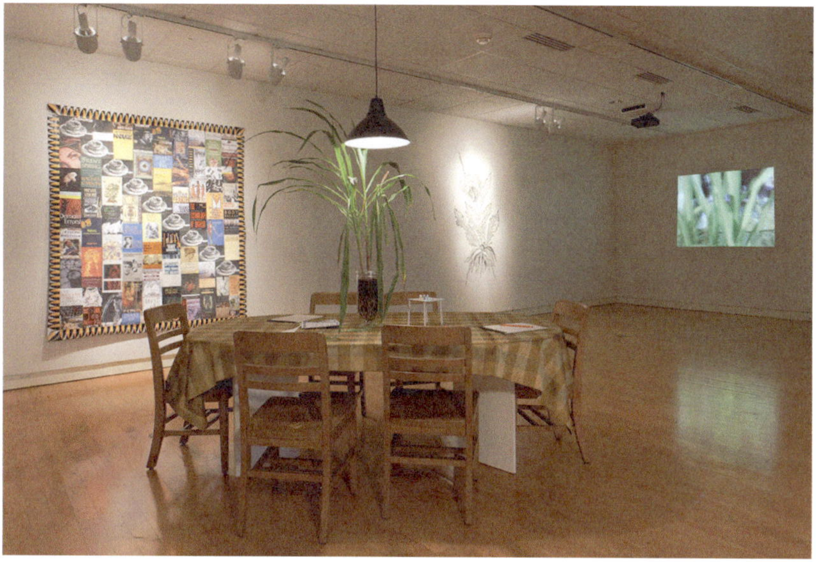

Fig. 6.30 A second wing of the gallery provided a larger gathering space for visitors to *Feminist Matter(s): Propositions and Undoings*. A large paper quilt comprised of book cover designs subverted the traditional form of the bibliography. The projected video showed our attempts at growing a patch of heritage Chiapan corn in a Pittsburgh backyard (subRosa 2011).

embedded in each setting—and to share their responses in the space. We intended to evoke intimate versions of the sometimes-improvised lab work benches, work spaces, and kitchens, in which many women scientists (and artists) did their first important work. In a different part of the gallery, we provided space for communal gatherings around a larger table, against the backdrop of a bibliographic book quilt (Fig. 6.30), composed from a patchwork of book cover designs from works inspirational to subRosa's practice.

Keywords: agriculture, domestic spaces, eugenics, feminism, feminist history of science, gentrification, installation art

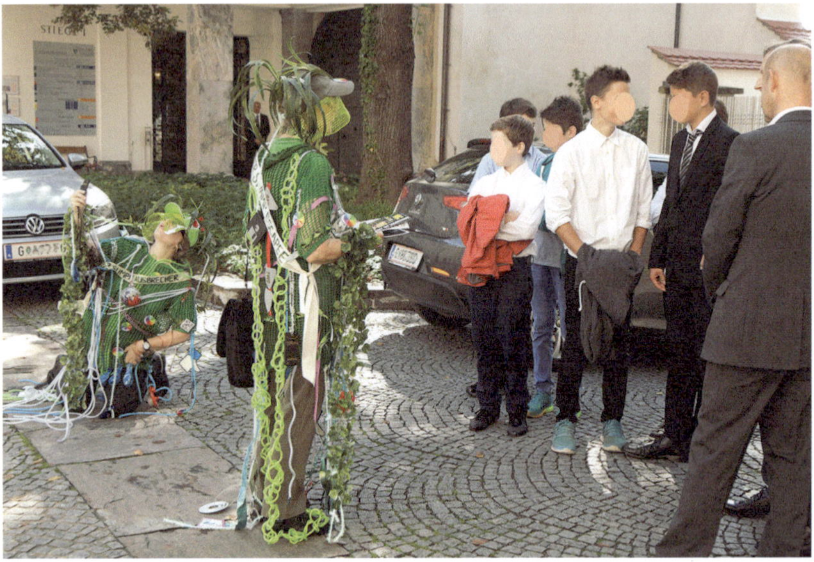

Fig. 6.31 subRosa's Cyberswamp Escapees roam the historic center of Graz, asking for help finding "Real Life." [auto plates and faces obscured for privacy] (subRosa 2014).

Down with Self-management! Re-booting Ourselves as Feminist Servers (2014)

Two ghostly figures drift slowly through the crowded historic town center of Graz, Austria, wearing costumes that appear to be dripping with bright green seaweed, flotsam, and jetsam (Fig. 6.31). A closer look reveals odd details: dangling social media icons, broken shards of old technology, and white sashes that read "Cyberswamp Escapee" in English and German. The pair drift towards strangers and ask for help understanding "Real Life" and distribute flyers with an invitation (printed in English, German, and Turkish) to visit a gallery location for "Kaffee und Kuchen" (traditional afternoon coffee, cake, and conversation). Graz is home to a substantial population of immigrants from Turkey and international war refugees from Syria, Iraq, Central Europe, and elsewhere. We encountered many tourists, groups of schoolchildren, senior citizens, religious workers, and immigrant vendors while strolling the streets.

Our costumes and interactions (which were conducted in German and English) sought to address the rapid development of digitized life and work, for example, techniques such as "life hacking," "social networking," "crowd-sourcing," "lean production," and "self-management". We wanted to hear from others what it feels like to live in this moment of compulsory self-management, self-marketing, and the enforced enhancement, upgrading, and "connectivity" of all aspects of our lives, including our reproductive capacities, and cradle-to-grave bodily optimizations.

subRosa asks: How do these ever-escalating demands depress our creativity and curtail our desire to work in common with others in RL? Can we help each other to resist drowning in the swamps of capitalist self-entrepreneurship by reimagining ourselves as a Collective-as-Server to support solidarity and collective action, and make common cause with the many different struggles for environmental, economic, social, and political justice?

Documentation and notes about our interactions were added to an experimental, mural-sized map of the city that was left in the gallery along with our costumes for the duration of the exhibition, which was commissioned for the *Steirischer Herbst* festival of new art and the *Ministry of Hacking* exhibition at ESC Median Kunst Labor (Graz, Austria 2014).

Keywords: cyberfeminism, experimental cartography, feminism, gentrification, hacking, installation art, performance art, social practice, tactical media, wearables

Gentrification U: A subRosa manifesta for Denver (2019)

Denver's surprising and layered history of tent communities and tent manufacturing was uncovered by subRosa in a large-scale annotated wall map of the neighborhoods surrounding the gallery in central Denver (Fig. 6.32). Our discoveries sought to raise questions about the role of art spaces and universities in the formation and development of cities and engage the higher education community and gallery visitors with issues of gentrification, precarious labor, and new forms of digitizing and monetizing of life and death that are impacted by the footprint of the university and gallery. The city flag for Denver features a stylized motif of twin mountain peaks that can also be interpreted as a pair of tents, and both symbols are enshrined in the architecture of the airport. We seized on this connection when exploring Denver's complex history of tent-like indigenous

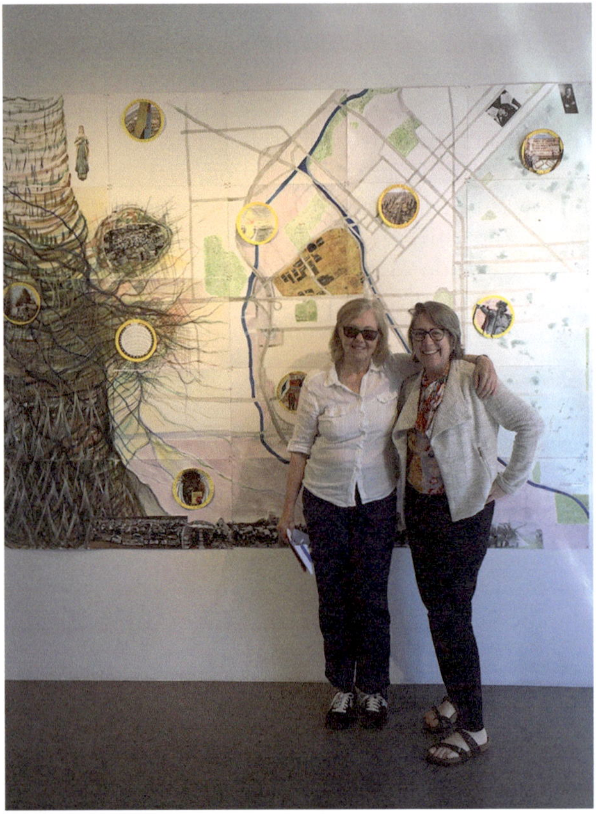

Fig. 6.32 *Gentrification U* was a large experimental map of downtown Denver, exploring its history of tentmaking, tent dwelling, and gentrification. The map was hand-drawn in ink and watercolor, on a matrix of illustration boards, with collaged elements. Following the show opening, subRosa members led two tours of the neighborhoods depicted in the map (subRosa 2019).

domiciles (the teepee), tent manufacturing (for leisure and military applications), and tent-based grassroots occupations and uprisings such as the Ludlow Massacre, Occupy Wall Street/Denver, and a growing population of unhoused individuals in the current wave of overpriced housing.

Gentrification U was drawn and collaged on a matrix of illustration boards so that it could be easily shipped, installed, and stored. Like our earlier *Cell Track* piece, there is an overlay of large discs, in this case functioning to add parallel or counter-narratives to some of the project's themes. In addition to the gallery exhibition, subRosa led a walking tour in the neighborhood surrounding the gallery, drawing attention to connections we made when doing research for this project.

Part of *Collectivism*, curated by Cecily Cullen, Tsehai Johnson, and Deanne Pytlinski, Center for Visual Art at MSU (Denver 2019).

Keywords: experimental cartography, feminism, gentrification, installation art

Index of Subjects

A
Abortion Joy
cyberfeminsim 27, 135, 136
Misoprostol 132, 133, 134
reproductive mutual aid 27, 107, 131, 132, 135, 136
subRosa 27, 107, 134, 135
tactical media 134
Women on Waves 135
Aging
cyberfeminism 1, 2, 13
digital culture 44, 182, 186, 187
memory 27, 153, 154
menopause 27, 159
refusal 117
seniors 153, 159, 160
technology 2, 5, 13, 27, 37, 153, 159, 186, 189, 197
Alchemy
alchemical emblem 120, 121, 180
alchemical reproduction 121, 130
alchemist 121, 123, 124, 125
ARTs 27, 108, 109, 121, 123, 124, 130, 135, 136
Black women 124, 126, 131
cyberfeminism 109, 121, 123, 124, 136
DIY Homunculi 105, 117, 133
Empire 123, 125
The Alchemy of Empire 125
transmutation 124
Zieglerin, Anna Maria 121, 123, 124, 125, 126, 136
Artificial Intelligence
bullshit 144
citizenship 50, 81, 144
consent 49, 55
cyberfeminism 2, 31, 121
durational art 26, 49
labor 2, 9, 31, 91, 145
lies 1, 145
privacy 45, 48, 55, 91

H. Willis et al., *Night Sweats: Cyberfeminist Practices*, Palgrave BioArt,
https://doi.org/10.1007/978-3-032-06955-9

refugees 48
Standard form Contracts 37
warfare 50
Assisted Reproductive Technologies [ARTs]
alchemy 27, 107, 108, 109, 121, 123, 124, 130, 135, 136
art 1, 2, 8, 39, 74, 75, 109
bioart 1, 8, 113
biopolitics 27, 124, 135, 136, 137, 138
birthing persons 9, 108, 124, 130, 131
Black women 124, 125, 126, 128, 129, 130, 131
compassionate inquiry 27, 107, 108, 109, 124, 130, 131, 135, 136
cyberfeminism 1, 2, 10, 39, 108, 109, 123, 135, 136, 143
eugenics 9, 39, 126, 137, 175
feminism 9, 39, 73, 109, 112, 122, 170
reproductive mutual aid 109
subRosa 1, 9, 39, 64, 107, 109, 113, 136, 142, 170
tactical media 1, 2

B

BioArt
biopolitics 7, 18, 24, 25, 124, 145
subRosa 1, 2, 7, 8, 10, 11, 12, 18, 24, 32, 62, 81, 94, 113, 124, 143, 146
tactical media 1, 2, 7, 12, 18, 24, 62, 81
Bioethics 10, 27
Biometrics
consent 47, 49, 50, 56
data collection 47, 50
privacy 45, 48, 53, 55, 96
refugees 48
surveillance 43, 45, 96, 97
user journey 45
warfare 50
Biopolitics
abortion joy 27, 133, 135
cyberfeminism 7, 12, 18, 20, 24, 25, 26, 117, 124, 135, 136
tactical media 12, 18, 24, 26
wetware 7, 21, 24
Biotechnology
ARTs 2, 108, 127, 137
bioart 2, 7, 31, 62, 94, 139
cyberfeminism 2, 4, 20, 26, 32, 61, 62, 109
Birthing persons 6, 17
Black Kinship, and Reproductive Mutual Aid 107
Black Women
Black kinship 107, 129
DIY Homunculi 107, 119, 136
reproductive mutual aid 107, 126, 128, 131, 135
transmutation 124, 125, 126

C

Christian Embryo Adoption 130
Close Readings of Terms and Conditions 56
Compassionate Inquiry 27, 107, 108, 109, 124, 130, 131, 135, 136
Consent
anxiety 52, 56
biometrics 47, 49, 50, 55, 56
citizenship 37, 50
privacy 36, 37, 46, 52, 55
shrink-wrap agreements 49
standard form contracts 37
terms and conditions 37, 46, 47, 49, 50, 55, 56
theatre 12
UX/UI 38

Consent Fatigue 52
Consent Theatre 52
Cyberfeminism
AI 2, 31, 121
alchemy 114, 121, 123, 124, 136
ARTs 1, 2, 10, 39, 108, 109, 123, 135, 136, 143
bioart 1, 2, 10, 11, 12, 18, 24, 32, 62, 94
biopolitics 12, 18, 20, 24, 25, 26, 117, 124, 135, 136
Cyberfeminism Index 2
Domain Errors! Cyberfeminist Practices 26, 181
Glitch Feminism
Net art, and Net culture 12, 31
performance art 31
racism 10, 64, 66, 71
satire 15, 81
subRosa 1, 2, 31, 39, 61, 62, 64, 66, 81, 116, 124, 135, 136, 143, 146, 173, 181
tactical media 1, 2, 3, 4, 7, 12, 18, 24, 26, 62
typography 15, 26, 31, 32, 39
Cybersecurity
AI 52
biometrics 52, 53
consent 52, 55
privacy 52, 53
refugees 48, 200
warfare 14
Cyborgs
aging 13, 27, 117, 159
alchemy 123
subRosa 13, 14, 27, 81, 82, 93, 160

D

DIY Homunculi, and plant-assisted reproduction, pARTs, and alchemy 107
DIY science
abortion joy 132
citizen science 136
DIY Homunculi 107, 117, 118, 119, 120, 136, 143
reproductive mutual aid 27, 107
tactical media 22
Domain Errors! Cyberfeminist Practices
cyberfeminist legacies
DE&I backlash 62, 66, 67
gender and sexuality panics
LGBTQIA+ backlash 62, 73, 116
pedagogy 62
Durational Art
"terms and conditions" 49
close reading of terms and conditions 56
fine print 39, 40, 41, 50, 156
opting out 26, 33
typography 26

E

Ectogenetics
AI 10
ARTs 9
cyberfeminism 10
desire 10, 17
wombs 9, 10, 17
Eugenics
Black feminism 9, 25, 39, 135, 170, 175, 184, 185
Black women 126, 139, 170, 175
cyberfeminism 9, 25, 39, 135
genetic determinism 25, 138
subRosa 8, 9, 25, 39, 126, 135, 137, 138, 139, 163, 170, 185

F

Feminist Art 7, 12, 32, 81, 90, 159
Feminist History of Science 12, 198

Feminist Pedagogy
alchemy 26
Domain Errors! Cyberfeminist Practices 3, 26
performance art 65
subRosa 26, 61, 62, 143, 199
Fine Print
consent 37, 50
dark patterns 38
forward looking statements 39, 41
shrink-wrap agreements 50
sludge 38
television advertising 42, 158
typographic scale 35

G
Genetic Screening
ARTs 127
Black women 126
genetic determinism 138
preimplantation 127
subRosa 126

I
In Vitro Fertilization [IVF] 3, 73, 76, 78, 79, 80, 127, 128, 129, 130, 131, 132, 142, 171, 172
Indecision Trees
DIY Homunculi 107, 136, 143
Luddites 137, 147
pARTs 107, 143
subRosa 107, 136, 137, 138, 139, 142, 143, 146

M
Magical Life 26, 27, 108
Matrixial Technologies
cyberfeminsim 4
ectogenetics 9
wombs 9, 10
Matrixial Vase
DIY Homunculi 118, 119
pARTs 118, 119
reproductive mutual aid 118
Media Studies
feminism 11
Gutenberg's Galaxy 35
pharmaceutical advertisements 42, 173
the global village 26, 33, 48
Tradwives 43
typography 26, 33, 34, 36, 41
Memory
aging 27, 153, 154, 156
AI 71, 153, 154, 155, 162
ars memoriae 50
brain-computer interfaces 43, 48
gaslighting
memorization 50, 154, 155
Menopause 27, 62, 81, 82, 159

O
Obadike, Mendi and Keith
Net art 15
racism 15
Opting Out
biometrics 53, 55
citizenship 24, 33, 55
durational art 26, 33
Luddites 147
refusal 24
terms and conditions 26, 33, 53, 55

P
pARTs (Plant Assisted Reproductive Technologies)
DIY Homunculi 107, 118, 119, 143
Indecision Trees 107, 143

Reproductive Justice 119
Performance Art
subRosa 113, 114, 116, 190, 192
Tsang, James (Wu) 110, 111, 113, 114, 115, 116, 190
Pregnancy
abortion joy 77, 79, 133, 135
ARTs 6, 10, 73, 77, 109, 127
Black women 120
ectogenetic desire 10, 17, 109
Matrixial vase 9, 119
mutual aid 76, 109, 127, 132, 133, 134
pARTs 120
reproductive justice 9, 120, 135
SmartMom 13, 15, 16, 17, 18, 43, 157, 164
subRosa 6, 10, 13, 15, 16, 17, 18, 43, 109, 120, 134, 165, 184
surrogacy 10, 14, 15, 73
surveillance 14, 16, 17, 43, 157
womb 10, 17, 127
Privacy Notices 36, 37, 38

R
Racism
ARTs 65, 170
cyberfeminism 10, 66
Net art 12
new media 10, 66
tactical media 12
Refusal
Black kinship 107
genetic 146
terms and conditions 39
undoing gender 27
Reproductive Justice 9, 12, 39, 62, 76, 95, 98, 113, 119, 131, 133, 134, 135, 136
Reproductive Mutual Aid
abortion joy 132
ARTs 126, 127, 128
Black women 126, 128, 129, 131
IVF 127, 128, 129
pARTs 107

S
Sex and Gender
intersex people 110
LGBTQIA+ 22, 73
subRosa 2
Trans people 118
Tsang, James (Wu) 110
undoing gender 113
useless gender 115
Shrink-wrap Agreements
consent
dark patterns 54
sludge 54
user journey 39
Standard Form Contracts
"opting out" 33
consent 37
dark patterns 38
durational art 39
refusal 39
sludge 38
typography 37
UX/UI design 36
subRosa
bioart 1, 2, 7, 8, 10, 11, 12, 18, 24, 32, 62, 81, 94, 113, 124, 143, 146
chronology of projects 27, 31, 163
collaboration 1, 5, 43, 56, 109, 110, 115, 116, 121, 147, 163, 164, 170, 181, 183, 190, 192
cyberfeminism 1, 2, 4, 31, 39, 61, 62, 64, 66, 81, 116, 124, 135, 136, 143, 146, 173, 181
eugenics 8, 9, 25, 39, 126, 135, 137, 138, 139, 163, 170, 185

feminist art 12
graphic design 15
humor 22
intellectual property 25
Net art 12
new media 7
origins of the group 48
performance art 8
tactical biopolitics 7, 18
tactical media 1, 2, 12
visual rhetoric 21
working methods of the group 2
workshops 5
Surrogacy
Black women 14, 15, 73, 77, 79, 97, 124, 128, 131
international fertility chains 123, 139
post-colonialism 10
Surveillance
biometrics 43, 45, 48, 96, 97
birth 14, 78, 127
brain-computer interfaces 43, 48
citizenship 14, 37, 81
consent 37, 47
data collection 47
extractive capitalism 2, 34
pregnancy 13, 14, 16, 43, 79, 127
privacy 37, 45, 48, 91, 96
refugees 48
the Biometric Assemblage 48
Tradwives 43
warfare 14
wearables 14, 43

T

Tactical Biopolitics
eugenics 25, 135, 138
subRosa 7, 18, 20, 24, 26, 117, 138, 146
Tactical Biopolitics (book) 7, 18
tactical media 18, 24, 25, 26, 135
Tactical Media
"strengthened and weakened" 24
bioart 1, 2, 7, 12, 18, 24, 62, 81
cyberfeminism 1, 2, 3, 4, 7, 12, 18, 24, 26, 62
subRosa 1, 2, 3, 4, 8, 12, 17, 18, 22, 24, 26, 62, 134, 165, 181
Tactical Biopolitics 135
typography 26, 35
Trans People
backlash 2, 75, 77, 80
gender 73, 74, 75, 85, 92, 109, 113
intersex people 110, 113
LGBTQIA+ 73, 75, 113, 116
sexuality and gender panics 1, 74, 75, 116
subRosa 2, 9, 109, 113, 116, 143
Tsang, James (Wu) 110, 111, 113, 116, 119
undoing gender 113
useless gender 115
Transmutation
alchemy 124
Black women 124, 125, 126
racism 124, 136
Typography
"fine print" 37, 39, 40, 41, 50
"terms and conditions" 31, 32, 34, 35, 36, 37, 39, 41, 51, 52, 56
"Typographic Man" 33
dark patterns 38, 56
fonts 34
graphic design 32, 34, 41
Gutenberg, Johannes 33, 34, 35, 41
McLuhan, Marshall 26, 31, 33, 34, 41
shrink-wrap agreements 50, 51, 52, 53
sludge 56
typographic norms 35
typographic scale 34, 35, 36

user pain points 38, 39
UX/UI design 38
visual rhetoric 32

U

Undercurrents
cyberfeminism 4, 10, 11
online communities 156
racism 10, 11
the post-colonial 4, 10, 11
User Experience Design and User Interface Design [UX/UI]
consent 37
cyberfeminism 32
dark patterns 38, 54, 55, 56
pain points 38, 39, 49
sludge 54, 55
typography 39

V

Vulvas
interactive art 97
subRosa 170

W

Weakened, and strengthened
biopolitics 18
tactical media 18
Wetware
subRosa 7, 21, 24, 136
tactical biopolitics 7, 18, 24, 25, 26
Women
aging 159
ARTs 1, 5, 9, 27
backlash 2, 62, 67, 68, 72
collaboration 5, 9, 43, 69, 99
cyberfeminism 1
gender and sexuality 1, 27, 68, 72
IVF 3, 73, 76, 78, 79, 80, 127, 128, 129, 130, 131, 132, 142
media studies 26
pedagogy 62, 198
race 10, 67, 100
reproductive mutual aid 27, 107, 108, 126, 127, 128, 129, 130, 131, 132, 135, 136
tactical media 1, 2, 12, 18, 19
weakened and strengthened 18, 23, 24
Women on Waves
abortion joy 78, 135
international solidarity 135
reproductive mutual aid 134, 135
Writing Systems
alphabets 33, 35, 56, 154
ars memoriae 50
automation 33, 34
calligraphy 34
handwriting 33, 46, 155
memory 27, 71, 154, 155, 162
movable type 31, 33, 35, 36
typography 15, 26, 31, 32, 33, 34, 36, 37, 39, 41, 52, 56

Z

Zieglerin, Anna Maria
alchemy 121, 123, 124, 125, 126, 136
magical life 107, 108, 114, 145
pARTs 121, 123

FSC
www.fsc.org
MIX
Papier aus verantwortungsvollen Quellen
Paper from responsible sources
FSC® C105338

If you have any concerns about our products,
you can contact us on
ProductSafety@springernature.com

In case Publisher is established outside the EU,
the EU authorized representative is:
Springer Nature Customer Service Center GmbH
Europaplatz 3, 69115 Heidelberg, Germany

Printed by Libri Plureos GmbH
in Hamburg, Germany